FLORA OF THE GUIANAS

Edited by

S. Mota de Oliveira and M.J. Jansen-Jacobs

Series A: Phanerogams

Fascicle 31

40. DILLENIACEAE
(G.H. Aymard C. & C.L. Kelloff)

117. VITACEAE
(J.A. Lombardi)

131. MELIACEAE
(T.D. Pennington & N. Biggs)

2016
Royal Botanic Gardens, Kew

First published in 2016 by
Royal Botanic Gardens, Kew,
Richmond, Surrey, TW9 3AB, UK
www.kew.org

Distributed on behalf of the Royal Botanic Gardens, Kew in North America by the University of Chicago Press, 1427 East 60th Street, Chicago, IL 60637, USA

ISBN 978-1-84246-465-6
e-ISBN 978-1-84246-623-0

British Library Cataloguing in Publication Data
A catalogue record for this book is available from the British Library

Typesetting by Christine Beard
Kew Publishing
Royal Botanic Gardens, Kew

Printed in the UK by Marston Book Services Ltd
Printed in the USA by The University of Chicago Press

For information or to purchase all Kew titles please visit
www.kewbooks.com or email publishing@kew.org

Kew's mission is to be the global resource in plant and fungal knowledge and the world's leading botanic garden.

Kew receives half of its running costs from Government through the Department for Environment, Food and Rural Affairs (Defra). All other funding needed to support Kew's vital work comes from members, foundations, donors and commercial activities including book sales.

Contents

The Flora of the Guianas

is a modern, critical and illustrated Flora of Guyana, Suriname, and French Guiana designed to treat Phanerogams as well as Cryptogams of the area.

Contents: Publication takes place in fascicles, each treating a single family, or a group of related families, in the following series: A: Phanerogams; B: Ferns and Fern allies; C: Bryophytes; D: Algae; and E: Fungi and Lichens. A list of numbered families in taxonomic order has been established for the Phanerogams.
Publication of fascicles will take place when available.
In the Supplementary series other relevant information concerning the plant collections from the Guianas appears, like indexes of plant collectors.

The Flora will, in general, follow the format of other Floras such as the *Flora of Ecuador* and *Flora Neotropica*. The treatments will provide fundamental and applied information; it will cover, when possible, wood anatomy, chemical analysis, economic uses, vernacular names, and data on endangered species.

ORGANIZATION: The Flora is a co-operative project of: Botanischer Garten und Botanisches Museum Berlin-Dahlem, *Berlin*; Institut de Recherche pour le Développement, IRD, Centre de Cayenne, *Cayenne*; Department of Biology, University of Guyana, *Georgetown*; Herbarium, Royal Botanic Gardens, *Kew*; New York Botanical Garden, *New York*; Nationaal Herbarium Suriname, *Paramaribo*; Muséum National d'Histoire Naturelle, *Paris*; Nationaal Herbarium Nederland, Naturalis Biodiversity Center, *Leiden*, and Department of Botany, Smithsonian Institution, *Washington, D.C.*

The Flora is edited by the Advisory Board: Executive Editors: S. MOTA DE OLIVEIRA, M.J. JANSEN-JACOBS, Leiden. Members: NILS KÖSTER, Berlin; P.G. DELPRETE, Cayenne; P. DA SILVA, Georgetown; E. LUCAS, Kew; T.R. VAN ANDEL, Leiden; B. TORKE, New York; D. TRAAG, Paramaribo; O. PONCY, Paris, and P. ACEVEDO RODRÍGUEZ, Washington, D.C.

PUBLICATION: The *Flora of the Guianas* is a publication of The Royal Botanic Gardens, Kew. The price of the fascicles will be determined by their size. Authors are requested to submit the manuscript in an electronic version; WORD and other major systems are acceptable.

INFORMATION: http://www.nationaalherbarium.nl/FoGWebsite/index.html

Editorial office (for correspondence on contributions, etc.):
S. Mota de Oliveira
M.J. Jansen-Jacobs
Nationaal Herbarium Nederland
Naturalis Biodiversity Center
P.O. Box 9517
2300 RA Leiden
The Netherlands
Email: sylvia.motadeoliveira@naturalis.nl

40. DILLENIACEAE

by

GERARDO A. AYMARD C.[1] & CAROL L. KELLOFF[2]

Lianas, shrubs, small trees with tortuous branches (*Curatella*), or evergreen trees up to 30 m (*Dillenia*), rarely herbs (not in the Guianas); lianas generally with stems > 5 cm diam., vascular bundles disposed in bands or concentric rings separated by abundant parenchyma (mostly in *Doliocarpus, Neodillenia* and *Pinzona*, several species of *Davilla* and *Tetracera*). Stipules absent or, if present, soon deciduous, infrequently winged and adnate to petiole. Leaves alternate, rarely opposite (not in the Guianas) simple, spirally arranged; leaf blade entire or dentate, usually with numerous parallel lateral veins, frequently coriaceous and scabrous, glabrous or pubescent with sclerified or silicified simple (*Curatella*) or fasciculate trichomes, occasionally stellate (*Tetracera*). Inflorescence terminal, axillary, or ramiflorous, paniculate, racemose, cymose, fasciculate (*Doliocarpus*), or flowers solitary. Flowers actinomorphic or rarely zygomorphic, hypogynous, bisexual or androdioecious (*Tetracera*); sepals 2-7(-14, in *Tetracera*), frequently 5, imbricate, sometimes unequal, the 2 inner longer and covering the fruit (*Davilla*), cucullate and imbricate for most of length, 2 larger than others in *Davilla*; petals 2-7, free or connate at base, imbricate, often crumpled in bud, deciduous, usually thin and delicate, yellow, white, or rarely pinkish; stamens usually numerous (20-500), centrifugal, free and persistent, connective broadened and sometimes thickened or connective linear, anthers basifixed, opening by longitudinal slits or apical pores; carpels 1-20, free or connate along ventral side (*Curatella* and *Pinzona*); placentation parietal to basal, ovules 1 or 2 or numerous in each carpel, apotropous or anatropous; styles free, usually slender and elongate, stigmas terminal, capitate or peltate. Fruits dry and dehiscent follicles or capsule, or berry-like, surrounded by 2 accrescent sepals (*Davilla*) or by all sepals (*Neodillenia*); seeds often arillate, endosperm abundant, carnose, embryo small, straight.

Distribution: widely distributed especially in tropical regions, ca. 300 species in 12 genera; in the Neotropical area ca. 90 species in 6 genera; in the Guianas 27 species in 6 genera.

---

[1] UNELLEZ-Guanare, Programa de Ciencias del Agro y el Mar, Herbario Universitario (PORT), Mesa de Cavacas. estado Portuguesa, Venezuela, 3350.
[2] United States National Herbarium, Department of Botany, NMNH, MRC-166, Smithsonian Institution, P.O. Box 37012, Washington, DC 20013-7012, U.S.A.

LITERATURE

Aymard, G. 2015. Novelties in Dilleniaceae from Ecuador. Harvard Papers in Botany 20(2): 209–212.

Aymard, G. 2010 onwards. Dilleniaceae. In: Neotropikey. Royal Botanic Gardens, Kew. http://www.kew.org/science/tropamerica/neotropikey.htm

Aymard, G. 1998. Dilleniaceae. In J.A. Steyermark *et al.* (eds.), Flora of the Venezuelan Guayana 4: 671-684.

Aymard, G. & J. Miller. 1994. Dilleniaceae Novae Neotropicarum III. Sinopsis y Adiciones a las Dilleniaceae del Perú. Candollea 49(1): 169-182.

Aymard, G. & S.A. Mori. 2002. Dilleniaceae. In S.A. Mori *et al.*, Guide to the Vascular Plants of Central French Guiana. Part 2. Mem. New York Bot. Gard. 76(2): 247-251.

Fraga, N.C. 2012. Filogenia e revisão taxonômica de Davilla Vand. (Dilleniaceae). PhD Thesis, Universidade Federal de Minas Gerais, Instituto de Ciências Biológicas. Departamento de Botânica. Belo Horizonte, Brasil, 423 pp.

Fraga, N. C. & J. R. Stehmann. 2010. Novidades taxonômicas para Dilleniaceae Salisb. Brasileiras. Rodroguésia 61 (Sup.): 1-6.

Horn, J. W. 2009. Phylogenetics of Dilleniaceae using sequence data from plastid loci (rbcL, infA, rps4, rpl16 INFRON). Int. J. Plant Sci. 170(6): 794-813.

Horn, J. W. 2006. Dilleniaceae. Pp. 132-154, in Kubitzki, K. (ed.), The Families and Genera of Vascular Plants. Volume IX. Flowering Plants. Eudicots. Berberidopsidales, Buxales, Crossosomatales. Springer, Berlin

Jansen-Jacobs, M.J. 1986. Dilleniaceae. In A.L. Stoffers & L.C. Lindeman, Flora of Suriname, Additions and corrections to 3(1-2): 475-484.

Kubitzki, K. 1971. Doliocarpus, Davilla, und verwandte Gattungen (Dilleniaceae). Mitt. Bot. Staatssamml. München 9: 1-105.

Kubitzki, K. 2004. Dilleniaceae. In: Flowering Plants of the Neotropics. 128-130 Pp. N. P. Smith *et al.* (eds.). Princenton University Press, Princeton. NJ.

Lanjouw, J. & P.F. van Heerdt. 1941. Dilleniaceae. In A. Pulle, Flora of Suriname 3(1): 386-408. E. J. Brill, Leiden.

Todzia, C. & G. Aymard. 2013. Dilleniaceae. Flora Mesoamericana 2(1): 1-15. http://www.tropicos.org/docs/meso/dilleniaceae.pdf

KEY TO THE GENERA

1. Small trees, erect shrubs or suffrutescent, no more than 15 m tall 2
   Climber shrubs or woody lianas 5

2. Stems and branches tortuous; leaves with silicified trichomes. .1. *Curatella*
Stems and branches not tortuous; leaves without silicified trichomes 3

3. Flowers androdioecious; fruits follicles; aril deeply laciniate . .6. *Tetracera*
Flowers bisexual; fruits berries or capsules; aril entire 4

4. Inflorescences panicles or clustered cymes, the inner sepals rigid-coriaceous,
covering the fruit .... 2. *Davilla*
Inflorescences ramiflorous, in fascicles or racemes; all sepals membranaceous,
never covering the fruit .... 3. *Doliocarpus*

5. Sepals unequal in size, covering the fruit entirely 6
Sepals more or less equal in size, if unequal, this never rigid-coriaceous
when mature and never covering the fruit entirely................. 7

6. Flowers buds ca. 0.5 cm long, only the two inner sepals rigid-coriaceous and
covering the fruit; aril white........................... 2. *Davilla*
Flower buds 1-4 cm long, all sepals membranaceous covering the fruit; aril
red.. 4. *Neodillenia*

7. Flowers androdioecious; fruits follicles; aril deeply laciniate . .6. *Tetracera*
Flowers bisexual; fruits capsules or berries; aril entire 8

8. Inflorescences not ramiflorous; carpels 2, connate ventrally from base to
apex of the ovary; aril orange......................... 5. *Pinzona*
Inflorescences ramiflorous along the stems; carpel 1; aril white ....
.... 3. *Doliocarpus*

1. **CURATELLA** Loefl., Iter. Hispan., 229: 260. 1758; Jussieu, Gen: 282. 1789.
Type: *Curatella americana* L.

Shrubs or small trees. Stipules absent. Leaves simple, alternate, pinnately veined, dentate-undulate, coriaceous and scabrous, elliptic or ovate-elliptic, hairs stellate, apex obtuse or rounded-emarginate, base rounded; petioles 0.8-1.5 cm long. Inflorescence paniculate, axillary, rarely terminal, tomentose. Flowers bisexual, actinomorphic, calyx persistent; sepals 3-5, puberulous, imbricate; petals 3-4, imbricate, free, deciduous, glabrous, obovate; stamens numerous, free; filaments filiform; anthers oblong, dehiscing longitudinally; ovary superior, carpels 2, connate ventrally from the base up to half of the ovary length, unilocular, ovules 2; styles 2, free, filiform; stigma capitate. Fruit a capsule, fused at the base, dehiscent along the ventral and lateral sutures into 4 valves, red within; seeds 2 per locule, smooth, obovate, black, completely covered by a striate, white aril, more or less entire.

Distribution: A monotypic neotropical genus found from Mexico through Central America, the Antilles and South America to the Brazilian "cerrado".

4

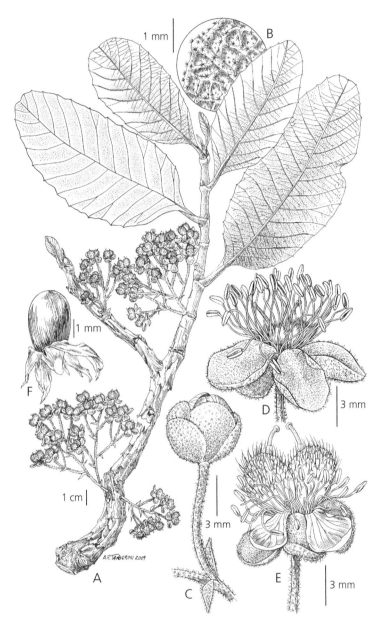

Fig. 1. *Curatella americana* L.: A, Habit; B, Detail of lower leaf blade with stellate hairs; C, Flower bud; D, Male flower; E, Female flower in fruit with petals attached; F, Seed with aril. Drawing by Alice Tangerini with permission of the Smithsonian Institution (A, E-F, McDowell & Tiwari 1840; C-D, Kelloff *et al.* 1159; B, Jansen-Jacobs *et al.* 2052).

LITERATURE

Todzia, C.A. & F.R. Barrie. 1991. Neotypification of Curatella americana L. (Dilleniaceae). Taxon 40: 488- 489.

1. **Curatella americana** L., Syst. Ed. 10:1079. 1759. Holotype: Venezuela: Loefling 260 (LINN, destroyed). Neotype: Venezuela, Guárico, around "San Juan de Los Morros", Williams & Alston 97 (neotype LL, isoneotypes F, M, US).   — Fig. 1

Tree, small (<15m), or shrub. Trunk short and branches tortuous, stunted. Bark thick, scabrous, gray-brown, covered by simple and stellate trichomes, barking off when mature. Stipules absent. Leaves stellate pubescent on both sides, elliptic or ovate-elliptic, 8-25 × 5-15 cm, apex obtuse or rounded-emarginate, base rounded; petioles reflexed-winged, canaliculate 0.8-1.5 cm long. Inflorescence 5-30 cm long, densely stellate-tomentose, pedicels 1-1.5 cm long, densely to sparsely stellate-pubescent. Flowers actinomorphic, calyx persistent; sepals 3-5, subequal, imbricate, ovate to wide elliptic, 5-7 × 3-5 mm, stellate-pubescent in both surfaces; petals 3-4, obovate, free, imbricate, deciduous, white, glabrous, 5-7 × 3-5 mm; stamens 50-100, free, persistent around the fruit; filaments 5-7 mm long, glabrous, filiform, dilated at the apex; anthers oblong, 0.5-0.6 mm long, dehiscing longitudinally; ovary superior, densely covered with white hispid hairs; carpels 2, connate ventrally from the base up to half the length of the ovary, 1-loculate and with 2 ovules; styles 2, free, filiform, glabrous; stigma small, capitate. Fruit capsule didymous, globose, pilose, fused basally, dehiscent along the ventral and lateral sutures into 4 valves, red within, carpels each with 2 seeds, rarely 1; seeds 2 per locule, smooth, obovate, black, ca. 4 mm long, ca. 2 mm wide, completely covered by a striate, white aril, more or less entire, nearly enveloping the seed at maturity, the embryo small.

Distribution: This is the most characteristic tree species of the macrothermic shrubby savannas in the neotropics. It is found between 0 to 1000m asl. Tree in open savannas, savanna margins and desciduous forests. 120 collections studied of which 30 collections from the Guianas. (GU: 19; SU: 5; FG: 6).

Selected specimens: Guyana: Rupununi R., Jenman 5580 (NY); Potaro R., Maguire & Fanshawe 23006 (M, NY). Suriname: Sipaliwini, Olderburger *et al.* 300 (NY). French Guiana: St. Jean Maroni, Benoist 842 (P); Ile de Cayenne, Oldeman 3936 (CAY, NY).

Vernacular name: Guyana: emiari (Wapishana), mimilli, sand paper tree. Suriname: boesi bisi, boesi kasjoe, bosch kasjoe, koela-ta, koerata mimillie, sabana kadjoe, sabana kasjoe, taja tatai.

Phenology: Flowers reported from December till May, fruits from July till November.

Uses: The leaves used to smooth bows, etc., and mild abrasive.

2. **DAVILLA** Vand., Fl. Lusit. Bras. Spec. 35: 14. 1788.
   Type: *D. rugosa* Poir.

Small climbing shrubs or woody lianas, branches and branchlets glabrous or with hirsute-ferrugineous pubescence or sparsely pilose. Leaves simple, alternate, chartaceous, coriaceous to subcoriaceous, lanceolate, ovate, suborbiculate, elliptic or oblong, petiolate, glabrous or with simple hairs; petioles winged to narrowly winged. Inflorescence paniculate, terminal or lateral, glabrous or ferrugineous-villous pubescent. Flowers bisexual; sepals 5, unequal in size, the two inner ones elliptic, 4-6 mm long, rigid-coriaceous when mature and covering the fruit, the three outer ones suborbiculate, 2-5 mm long; petals 3-6, obovate, deciduous, glabrous, yellow; stamens numerous, 50-450; carpels 1-2, free; ovary superior, glabrous, 1-locular, ovules 2 per locule; style sublateral; stigma peltate. Fruit a capsule, orange or yellow; seeds 1-2, arillate.

Distribution: Neotropical genus, with approximately 30 species. Known from Mexico, Central America, the Antilles, Venezuela, Colombia, Peru, Bolivia, Guianas, Brazil and Paraguay. In the Guianas 5 species recorded, this treatment includes one more species, expected to occur in the Guianas.

The genus *Davilla* Vand. was formerly divided in two sections based on differences of the inner sepals: sect. *Davilla* having overlapping internal sepals, and sect. *Homalochleaena* Kub. with the internal sepals not imbricate, pressed against each other (Kubitzki, 1971). More recently, Fraga (2012), found additional support from morphological and molecular data for two new sections: *Dryadica* Fraga (stamens ≥ 50, filaments cilindrical), and *Complanata* Fraga, Smidt & Stehmann (stamens ≤ 50, filaments flat).

KEY TO THE SPECIES

1. Carpels 2; petioles fully winged 2
   Carpel 1; petioles narrowly winged or unwinged 3

2. Leaves elliptic, adaxial surface smooth, abaxial surface sparsely yellow pubescent, venation not lacunose-areolate, glabrescent when mature; petioles 4-6.5 cm long . . . . 1. *D. alata*
   Leaves elliptic to elliptic obovate, adaxial surface scabrous, abaxial surface densely ferrugineous pubescent, venation mostly lacunose-areolate; petioles 1-2 cm long . . . . 6. *D. steyermarkii*

3. Leaves and outer sepals glabrous or nearly so, venation eucamptodromous
   .... 4. *D. nitida*
   Leaves and outer sepals pilose, pubescent or scabrous, venation
   brochidodromous 4

4. Midrib on abaxial surface and inflorescence patent-pilose, ferrugineous,
   not scabrous .... 5. *D. rugosa* var. *rugosa*
   Midrib on abaxial surface and inflorescence adpressed-pubescent,
   scabrous.. 5

5. Leaves elliptic, suborbicular or oblong, abaxial surface glaucous; tertiary veins
   on abaxial surface mostly prominent, areole bullate .... 3. *D. lacunosa*
   Leaves elliptic to widely elliptic, abaxial surface not glaucous; tertiary veins
   on abaxial surface not prominent, areole not bullate .... 2. *D. kunthii*

1. **Davilla alata** (Vent.) Briq., Annuaire Conserv. Jard. Bot. Genève 4:
   217. 1900 – *Curatella alata* Vent., Choix Pl., 49. 1803. Type: French
   Guiana. Martin s.n. (holotype G, isotypes BM, P).

   Davilla vaginata Eichl. Fl. Bras. 13(1): 99. 1863. Type: French Guiana,
   Martin 32 (P).
   Davilla wormiaefolia Baill., Adansonia 6: 272. 1866. Syntypes: French
   Guiana: Martin s.n (P), Melinon 1864 (P, US).

Small tree or liana, young branches and inflorescence scabrous. Leaves
coriaceous, elliptic, 12-22.5(-29) × 8.5-12 cm, apex apiculate or acute,
base acute, petiolate, margin dentate, slightly revolute; leaf blade glabrous
adaxially, golden brown setose abaxially, with 13-17(-24) pairs of lateral veins,
venation craspedodromus; midrib and lateral veins impressed and glabrous
adaxially, prominent and setose abaxially; petiole 4.5-6.5 cm long, fully
winged, wings 5-6 mm wide extending almost to leaf blade, with appressed
strigose hairs. Inflorescence axillary, fascicles with 3-5 pedunculate flowers;
peduncles 1-1.5 cm long, appressed strigose hairs. Flowers whitish to yellow,
ca. 1.0 cm broad; petals 5, deeply emarginate; stamens ca. 150, 5-6 mm long;
sepals with golden-brown (reddish-brown) appressed strigose hairs on the
outside, glabrous within; outer sepals ca. 1.2-1.5 cm wide, depressed ovate,
inner sepals ca. 2.0 cm wide hemispherical with flat, ca. 2 mm wide margins,
enclosing the fruit; carpels 2, glabrous. Fruit ca. 2 cm wide obovate, 2 seeded;
seed dark, obovate-subreniform, 5-6 mm long.

This species is found in the Guianas; 21 collections studied, all from the
Guianas: (GU: 3; SU: 6; FG: 12).

Selected specimens: Guyana: Konashen-area, Essequibo R. at
Saparimo, Jansen-Jacobs 1797 (US, BBS); Black Cr. Groete Cr. Essequibo
R., Fanshawe 1790 [Forest Dept. 4526] (NY, K). Suriname: Brokopondo
Distr.; 6 km E of village Brownsweg, van Donselaar 3219 (K); Brokopondo

8

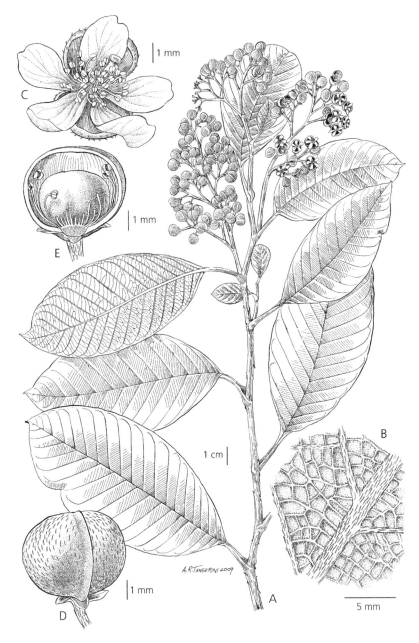

1 mm

1 mm

1 cm

A. R.Tangerini 2009

1 mm

5 mm

Fig. 2. *Davilla kunthii* A. St.-Hil.: A, Flowering branch; B, Lower leaf surface; C, Flower with carpels open; D, Fruit surrounded by both carpels; E, Fruit with one carpel removed. Drawing by Alice Tangerini with permission of the Smithsonian Institution (A-B, Acevedo & Tiwari 3258; C, Cowan & Soderstrom 2045; D-E, Pipoly *et al.* 11355).

Distr., Forest outside National Reserve Brownsberg, near Brownsweg, Lindeman 52 (NY, BBS). French Guiana: Maroni, Melinon s.n. (K, US); Crique Plomb, Bassin du Sinnamary, Loubry 1758 (US).

Vernacular names: Guyana: kabuduli (Ar), katuwai (Ak). French Guiana: asrika-Tite, dialoppoe.

2. **Davilla kunthii** A. St.-Hil., Pl. Usuel. Bras., 6. 1824 [1825]. Type: Venezuela, Humboldt 252 (P)   – Fig. 2

Tigarea aspera Aubl., Hist. Pl. Guiane 2: 918. 1775. French Guiana, Lamarck s.n (MO)
Davilla asperrima Splitg., Tijdschr. Nat. Geschied. 9: 95. 1842. Type: Suriname, Splitgerber 503 (U; isotype P, W).
Davilla surinamensis Miq., Linnaea 18: 611. 1844. Suriname, Focke 873 (U; isotype: K)

Woody liana, young branches scabrous, later bark peeling off in layers. Leaves coriaceous, elliptic to widely elliptic, 8-15(17) × 4.5-7(9) cm, apex acute to acuminate, base attenuate, leaf blade glabrous adaxially, apressed pubescent and scabrous abaxially, with 12-20(-23) pairs of lateral veins; midrib and lateral veins impressed adaxially, prominent abaxially; petioles glabrous to slightly pubescent, 1.5-2.4(3.4) cm long, narrowly winged. Inflorescence paniculate, in terminal axil, scabrous when young with many flowers; outer sepals widely elliptic to widely depressed ovate, ca. 1.0-1.2 mm, sparsely strigose, inner sepals hemispherical, orangish yellow to brown, ca. 5 mm in diameter, scabrous, with overlapping margins, ca. 1 mm of interior margin folded back under exterior sepal and enclosing the ovary; petals 5, yellow to yellow-orange, emarginate; stamens 40-50; style 1, peltate; ovary glabrous, 1 seeded.

Note: *Davilla kunthii* St. Hil. is the most polymorphic species of the genus. Fraga (2012) considered the latter a synonym of *D. nitida* (Vahl) Kub. However, in the Flora of the Guianas treatment, these taxa are treated as different species, following Kubitzki´s (1971) and Aymard´s (1998) concepts, based on the morphological features used in the descriptions, and on the current key.

Distribution: Mexico, Central America, Antilles, Venezuela, Colombia, Guyana, Suriname, French Guiana, Ecuador, Peru, Brazil, Bolivia; 137 collections studied, all from the Guianas: (GU: 68; SU: 30; FG: 39). Common in savannas, edges of forests, and disturbed areas.

Selected specimens: Guyana: U. Demerara-Berbice Region, Rockstone, in and about the village, Gleason 669 (US, NY, K); Barima-Waini Region, Waini R., NW District, de la Cruz 3595 (US, NY). Suriname: Road to

Matawaribo, ca 30 km S of Paramaribo, Heyde 702 (BBS); Lely Mts, SW plateaus covered by ferrobauxite, Lindeman *et al.* 265 (NY, K, BBS). French Guiana: Savane de Corossony, PK 111 de la route Cayenne-St. Laurent, Cremers 555 (US, P, CAY); Ile de Cayenne, sur la route de Raban, Oldeman 1255 (US, P, CAY)

Phenology: Flowering February to June

Vernacular names: Guyana: kabuduli (Ar), katuwai (Ak). Suriname: brandliaan, faya tati (Saramacca), sah-kuh-ti-to (Tirio), Kawtite, diatite (Sranan).

3. **Davilla lacunosa** Mart., Flora, 21 (2, Beibl 4): 49. 1838. Lectotype: Brazil, morro do Ernesto, near Cuiabá, Da Silva Manso 107 in Mart., Hb. Fl. Bras. 219 (BR) (designated by Kubitzki, 1971)[3]. Syntypes: Martius 219 (K, M, MO, NY, P).

Small tree to 3 m tall, liana or scandent shrub, young branches with appressed setose pubescence, older branches glabrous or nearly so, peeling off in layers. Leaves rigid coriaceous, elliptic or suborbicular or oblong, 6-14 × 3.5-8.5 cm, apex round sometimes emarginate, base petiolate rotundate acute, rarely rotundate or obtuse, margins entire, sometimes subsinuate, (sub) revolute; leaf blade with (8-) 12-19 pairs of lateral veins, venation brochidodromus; midrib with appressed strigose hairs, lateral and tertiary veins minute pubescent, impressed adaxially, glaucous and prominent abaxially, areola bullate; petioles semi-triangular, 1-2.5 cm long, sometimes with 0.5-1 mm recurved wings, minutely setose. Inflorescence paniculate, 5-10 cm long, minutely pubescent; bracts lanceolate-ovate, ca. 2 mm long; sepals warty with very little minutely scabrous and very short soft hairs on the outside; outer sepals orbicular, ca. 2-4 mm long; inner sepals ca. 6 mm long, elliptical with loosely appressed setose pubescence, enclosing the fruit; petals 5, yellowish-green; stamens 35-40, 3-4 mm long; carpel 1, glabrous, 1 seeded; seed globose, ca. 5 mm diameter.

Distribution: Brazil and French Guiana; 11 collections studied, all from French Guiana.

Selected specimens: French Guiana: Ile de Cayenne, cote Ouest sur al Route de Radan, Oldeman 1255B (US, P); Ile de Cayenne, sur la route de Rochambeau au km 8, estabissement Jalmar; Oldeman 3930B (NY); Route du Tour de l'Ile pres de la RN1, Feuillet 606 (P).

---

3 Note that the syntype collections in P and K were also annotated as "Lectotype" by Kubtizki, in 1969, but the collection in BR was the one lectotypified in his work (Kubitzki, 1971).

11

Vernacular names: Guyana: kabuduli (Ar), katuwai (Ak). French
Guiana: bopn-zete (P), kabudeli (Ar).

4. **Davilla nitida** (Vahl) Kubitzki, Mitt. Bot. Staatssamml. München
9: 95. 1971. – *Tetracera nitida* Vahl, Symb. Bot. 3: 70. 1794. Type:
French Guiana, von Rohr s.n. (C, photo M)

Tetracera multiflora DC., Syst. Nat. 1: 400. 1818 [1817]. Brazil, Para,
without collector (P)
Davilla multiflora (DC.) A. St.-Hil., Pl. Usuel. Bras., 5: 5, t. 22 1824 [1825].

Woody liana to scandent shrub, young branches scabrous, later bark peeling
off in layers. Leaves coriaceous, elliptic to ovate, 10-15.5 × 5-6.5 cm, apex
acute to acuminate, base obtuse to attenuate; leaf blade scabrous on both
sides (or sometimes smooth adaxially), with 10-14(-16) pairs of lateral
veins, venation eucamptodromous; midrib and lateral veins impressed
adaxially, prominent and strigose abaxially; petioles glabrous to slightly
pubescent, 1.0-2.5 cm long, narrowly winged. Inflorescence paniculate in
terminal axil, many flowered; outer sepals widely elliptic to widely ovate,
ca. 1.0-1.1 mm wide, glabrous or nearly so, inner sepals hemispherical,
orangish yellow, ca. 5-7 mm in diameter, minute scabrous to glabrous,
with overlapping margins, enclosing the ovary; petals 5, emarginate,
orange-yellow; stamens 30-50; style 1, peltate; ovary glabrous, 1 seeded.

Distribution: Mexico, Central America, Antilles, Venezuela, Colombia,
Guyana, Suriname, French Guiana, Ecuador, Peru, Brazil, Bolivia; 16
collections studied, all from the Guianas: (GU: 7; SU: 1; FG: 1).

Selected specimens: Guyana: U. Takutu-U. Essequibo Region;
Rupununi Northern Savanna, Yupukari, Goodland 950 (US); Potaro-
Siparuni Region, Upper Mazaruni River, S slope of Karowtipu, 460-
780 elev.; humid forest, Boom & Gopaul 7351. Suriname: Brownsberg,
Burean 6554 (NY, K). French Guiana: Lotissement Ames-Claires, Ile de
Cayenne, Hoff 5356 (US).

Vernacular names: Guyana: kabuduli (Ar), katuwai (Ak).

5. **Davilla rugosa** Poir., Encycl. (Lamarck) Suppl. 2: 457. 1812. Type:
Brazil, Dombey, s.n. (P-JU).

In the Guianas only: var. **rugosa**

Davilla pilosa Miq., Linnaea 19: 134. 1847 Type: Suriname, Kappler 1711
(holotype U, isotype G, P, S).

Scandent shrub, young branches ferruginous-hispid, older branches
scabrous later becoming glabrous or nearly so, bark peeling off in layers.

Leaves chartaceous or subcoriaceous, entire, elliptic, oblong or ovate, (4.5-)9.5-11 × 4-6.5 cm, apex acuminate to obtuse, base obtuse or acute, margin sub-revolute; leaf blade with (7-)9-13 pairs of lateral veins, venation brochidodromus (rarely craspedodromus and dentate), impressed with few short strigose hairs adaxially, prominent and ferrugineous pilose abaxially; petioles narrowly recurved winged, 1-1.2(-1.7) cm long, villose. Inflorescence paniculate in terminal axil, many flowered, ferruginous-villose; outer sepals ca. 1 mm long, inner sepals hemispherical, ca. 4 mm long, after anthesis increasing to 7-9 mm long, elliptical or sub-obovate, scabrous, with overlapping margins and enclosing the ovary; petals 5, yellow to yellowish-orange, round to short acuminate; stamens 40-50; style 1; ovary glabrous, 1 seeded.

Distribution: Neotropical; Central America, Cuba, Colombia, Venezuela, Guyana, Suriname, French Guiana, Ecuador, Peru, Brazil; 2 varieties, 1 in the Guianas. 31 collections studied, all from the Guianas: (GU: 5; SU: 13; FG: 13).

Selected specimens: Guyana: Suddie, Essequibo, Stockdale s.n. (BRG); Rupununi, north savanna, stand 40, Yupukari, Goodland 968 (NY). Suriname:"Great Granite Outcrop" in forest island, 5 km W of "Morro Grande" dome, Oldenburger ON456 (NY, BBS); without location, Wigman 50 (BBS). French Guiana: Karouabo, Sagot 17 (NY, K, P, BM); without location, Mélinon 85 (NY, P)

Phenology: Flowering in January and August. Fruiting in March and October.

Vernacular names: Guyana: kabuduli (Ar), katuwai (Ak). Suriname: zachte brandliaan.

6. **Davilla steyermarkii** Kubitzki, Mitt. Bot. Staatssamml. München 16: 501. 1980. Type: Venezuela, Bolívar: J.A. Steyermark *et al.* 117841 (holotype M, isotypes U, VEN).

Woody vine, with large branches, branchlets densely ferrugineous-villose, trichomes ferrugineous, sparsely strigose to glabrecent when mature. Leaves coriaceous, elliptic to elliptic-obovate, 3.5-12 × 2-6 cm, apex rotundate, sometimes emarginate, base obtuse or acute, narrow at the petiole, margins revolute, entire, scabrous; leaf blade strigose to glabrescent adaxially, strigose ferrugineous abaxially, indument more dense along the midrib and secondary veins, 7-17 lateral veins, venation eucamptodromous, convergent and linking to the margin, or linking close to the margin in the upper half, impressed adaxially, strongly reticulate abaxially; petioles connivent-alate to winged, 1-2 cm long, 3-5 mm wide, covered by ferrugineous trichomes. Inflorescence 5-8 cm long, densely ferrugineous pubescent; bracts deciduous, lanceolate, densely ferrugineous pubescent, ca. 4 cm, pedicels 4-10 mm long, ferrugineous; sepals suborbicular, coriaceous, densely ferrugineous-sericeous externally, glabrous internally, outer sepals 3-5 mm long, inner sepals elliptic, 7-15 mm long, ca. 1.7 cm long when mature; petals 3, yellow, 5-6 mm long, ovate, glabrous on both sides; stamens ca.180, filaments 5-6 mm long, glabrous; anthers ca. 0.8 mm long, glabrous; carpels 2, glabrous; style 9-11 mm long, glabrous; stigma capitate; seeds 2 per carpel, black, obovate-subreniform, ca. 5 mm long, laciniate envolving the seed.

Distribution: Expected in Guyana. Outside the flora area, *D. steyermarkii* Kub. is known from gallery forests and edges of savannas at 700 to 1100 m alt., over Roraima sandstone formations, in Southeastern Venezuela and Brazil border.

3. **DOLIOCARPUS** Rol., Kongl. Svenska Vetensk. Acad. Handl. 17: 260, 1756.
   Type: *Doliocarpus major* J.F. Gmel.

Lianas, vines, erect or scandent shrubs. Stems with vascular tissue arranged in bands or cencentric rings separated by abundant parenchyma. Stipules absent. Leaves smooth, dentate, sinuate or entire, alternate, pinnately veined, pubescent with simple trichomes, or glabrous; petioles winged, narrowly winged, or unwinged. Inflorescences axillary or ramiflorous along the stems, racemose, paniculate, fasciculate, glomerate, or the flowers solitary. Flowers bisexual; sepals 3-6, subequal, persistent; petals 2-6 (often 3-4), obovate, caducous; stamens numerous (20-80); filaments filiform; anthers with a broadened and thickened connective, rarely linear, longitudinally dehiscent; ovary 1-carpellar, 1-locular; ovules (1)2, basal, anatropous; style terminal, filiform; stigma peltate. Fruits capsules or berries; seeds 1 or 2, aril white, completely surrounding the seed.

Distribution: Neotropical genus with ca. 40 species, 12 of which occur in the Guianas. Known from Mexico, Central America, the Antilles, Venezuela, Colombia, Ecuador, Peru, Bolivia, Brazil and Paraguay.

The genus *Doliocarpus* Rolander was monographed by Kubitzki (1971), who divided it into two sections: section *Calinea* Eichler, characterized by having leaves with tertiary nerves subparallel (rarely reticulate), erect-flexuose filaments with introrse anthers at anthesis, and a glabrous or pilose ovary, and section *Doliocarpus* having leaves with tertiary nerves reticulate, reflexed filaments with anthers extrorse at anthesis, and ovary always pilose. Although the stamen character state (filaments at anthesis) represents a reliable morphological feature when assigning specimens of *Doliocarpus* to these two sections, it is usually the case that most specimens do not have flowers at anthesis, or have flowers lacking petals, many having instead young fruits with persistent sepals and stamens. Leaf venation, therefore, is perhaps the most valuable character to distinguish the two sections of this genus (Aymard, 2015).

KEY TO THE SPECIES

1. Suffrutescent to erect shrubs less than 4 m tall....................... 2
   Vines, climbing shrubs or woody lianas with stems more than 5 cm in circumference... 3

2. Leaves rigid-coriaceous, conduplicate, lateral veins running straight towards the margin or anastomosing at about 0. 2 mm from the margin; abaxial surface of leaves and fruit glabrous .... 12. *D. spraguei*
   Leaves coriaceous, not conduplicate; lateral veins always running straight towards the margin, midrid and lateral veins appressed pilose abaxially; fruit sparsely villose .... 11. *D. savannarum*

3. Tertiary venation reticulate, punctuate on abaxial surface 4
   Tertiary venation subparallel, not punctuate on abaxial surface......... 6

4. Leaves 2-7 × 2-5 cm; sepals 3, 2-4 mm long; ovary glabrous . . 4. *D. gracilis*
   Leaves 10-18 × 4-10 cm; sepals 4 or 5, 5-7 mm long; ovary pilose...... 5

5. Leaves lanceolate, lanceolate-elliptic or lanceolate-ovate, base rounded, sparsely strigulose abaxially; sepals obovate or elliptic; fruit 1-1.3 cm diameter .... 7. *D. major* subsp. *major*
   Leaves obovate-oblong or obovate-lanceolate, base cuneate, glabrous abaxially; sepals oblong; fruit 2-3 cm diameter......... 9. *D. paraensis*

6. Inflorescences fasciculate, glomerose, not racemose (peduncles 1 flowered) ....3. *D. dentatus*
   Inflorescences racemose (peduncles 2-8 flowered) 7

7. Inflorescences 1.5-4 cm long; fruit 1-1-1.5 cm diameter 8
   Inflorescences 0.2-1 cm long; fruit 0.5-0.8 cm diameter 9

8. Leaves obovate to obovate-oblong; margin entire or subsinuate; ovary glabrous, rarely sparsely pilose .... 5. *D. guianensis*
   Leaves obovate-lanceolate; margin mucronate-serrate; ovary villose .... 8. *D. multiflorus*

9. Leaves tomentose or densely apressed pilose abaxially; petioles 0.7-2 cm long; ovary pubescent....................................... 10
   Leaves glabrous, strigose to sparsely pilose abaxially; petioles 0.5-1 cm long; ovary glabrous....................................... 11

10. Leaves subcoriaceous, obovate, obovate-oblong, base cuneate; inflorescence
    produced from short shoots (brachyblastus), peduncle 0.5-3 mm long . . .
    . . . . 6. *D. macrocarpus*
    Leaves coriaceous, lanceolate, lanceolate-obovate, base acute; inflorescence
    produced from long shoots (dolichoblastus), peduncle 1-6 mm long . . . . .
    . . . . 1. *D. savannarum*

11. Leaves rigid-coriaceous, obovate or obovate-lanceolate, margins entire or
    subrepanded, glabrous abaxially 12
    Leaves charthaceous or coriaceous, lanceolate-elliptic or obovate-oblong,
    margins dentate, pilose abaxially 13

12. Leaves conduplicate; lateral veins running straight towards the margin or
    anastomosing at about 0. 2 mm from the margin; sepals 5 . . . .
    . . . . 12. *D. spraguei*
    Leaves not conduplicate; lateral veins running straight towards the margin;
    sepals 6 . . . . 10. *D. sagotianus*

13. Leaves coriaceous, lanceolate-elliptic, base attenuate; lateral veins 6-9;
    sepals sparsely pilose outside . . . .
    . . . . 2. *D. brevipedicellatus* subsp. *brevipedicellatus*
    Leaves chartaceous to subcoriaceous, obovate-oblong, base acute; lateral
    veins 10-16; sepals puberulent to glabrous outside . . . .
    . . . . 1. *D. amazonicus* subsp. *amazonicus*

1. **Doliocarpus amazonicus** Sleumer, Repert. Spec. Nov. Regni Veg. 39:
   44. 1935. Type: Peru. Loreto: Río Blanco, Tessman 3065 (holotype
   B†, G, not seen; isotype S, not seen)

Doliocarpus amazonicus Sleumer subsp. *amazonicus*

Climbing shrub to lianas, branches and branchlets apressed pilose,
glabrescent when mature. Leaves chartaceous or subcoriaceous, obovate-
oblong, 10-25 × 6-10 cm, apex wide rotundate, or acute, short acuminate,
base acute, margins serrate to crenate, mostly in the upper half; leaf blade
glabrous adaxially, pilosule along the midrib and secondary veins, apressed
pilose abaxially, with 10-16 pairs of lateral veins, running straight towards
the margin, impressed adaxially, prominent abaxially; petioles ca. 0.8 cm
long, shortly pubescent, narrowly alate. Inflorescences racemose, axillary,
0.5-1 cm long, densely compacted 2-4 flowered, peduncles ca. 3 mm
long, shortly pubescent. Flowers 3-5 mm broad; sepals 4-5, obovate or
elliptic, 2-4 mm long, puberulous outside, glabrous inside; petals not seen;
stamens ca. 35; carpels 1, glabrous, style ca. 1 mm long. Fruit 0.6-0.8 mm
in diam., glabrous, red, irregularly dehiscent; seed 1, with a white aril.

Distribution: Peru, Brazil (Amazonas, Mato Grosso, Rondônia) and
Bolivia. Lowland wet forests, up to 200 m alt.. 10 collections studied, 1
from the Guianas (SU: 1).

16

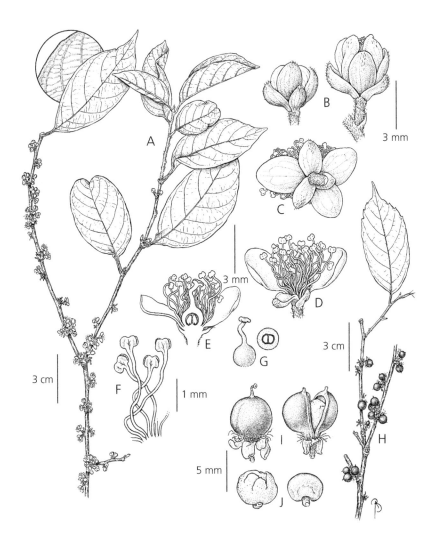

Fig. 3. *Doliocarpus brevipedicellatus* Garcke: A, Flowering branch; B, Floral buds. C,
Flower, basal view; D, Flower, side view; E, Flower, longitudinal section; F, Stamens;
G, Gynoecium, entire and cross section, showing two ovules; H, Fruiting branch; I,
Fruit, closed and open; J, Seed, with arillode and naked. From Mori, S.A. *et al.* 2003.
Vascular plants of central French Guiana. Mem. NYBG Vol. 76(2).

Selected specimens: Suriname. Sipaliwini: between Augusts creek and rim of arrowhead basin, 03° 54' N; 56° 11' W, Lohmann *et al.* 232 (BBS, MO, PORT).

2. **Doliocarpus brevipedicellatus** Garcke, Linnaea 22: 47. 1849. Type: Suriname, Jodensavanne, Kegel 1186 (GOET).

In the Guianas only: subsp.**brevipedicellatus**

–Fig. 3

Woody liana, young branches glabrous. Leaves coriaceous, lanceolate to elliptic, 6-15 × 2.5-7 cm, apex acuminate, base attenuate, margin coarsely dentate along ca. upper one-half; leaf blade strigose abaxially along veins, with 6-9 pairs of lateral veins, running straight towards the margin, the tertiary veins not raised, smooth; petioles 0.5-1 cm long, strigose, very narrowly alate. Inflorescences racemose, 0.5-1 cm long, fascicles of densely compacted 2-4 flowered racemes or flowers solitary; peduncles 1-3 mm long, sparsely pilose. Flowers 3-5 mm broad; sepals unequal, outer ones 1.5-2 mm long, inner ones 3-3,5 mm long, ovate-elliptic, sparsely pilose outside, glabrous within; petals 2-3 mm long, white; stamens 30-40; carpels 1, glabrous, style ca. 1 mm long. Fruit 5-6 mm in diam., glabrous, red, irregularly dehiscent; seed 1, with a white aril.

Distribution: Nicaragua, Panama, West Indies, Venezuela, Colombia, Peru and Brazil (Amazonas, Goiás, Maranhão, Mato Grosso, Para, Rondonia Roraima); Lowland wet forests, or riverbanks, up to 200 m alt. 70 collections studied, 15 from the Guianas (GU: 7; SU: 4; FG: 4).

Selected specimens: Guyana: Mazaruni Station, For. Dep. Brithish Guiana 2954 (K); Malali, Demerara R., de la Cruz 2673 (MO, US). Suriname: Brokopondo, van Donselaar 3125 (U). French Guiana: Montagne de Mahuri, Maas *et al.* 2212 (NY, U).

Phenology: Flowers reported from July to September, fruits from November, December and January.

Vernacular names: Guyana: kabaduli.

3. **Doliocarpus dentatus** (Aubl.) Standl., J. Wash. Acad. Sci. 15: 286. 1925. – *Tigarea dentata* Aubl. Hist. Pl. Guiane 2: 920. 1775. Type: French Guiana, Aublet s.n (BM).

Tetracera cuspidata G. Mey,, Prim. Fl. Esseq. 205. 1818. Type: Guyana, G. F. W. Meyer s/n (GOET)
Doliocarpus semidentatus Garcke, Linnaea 22: 48. 1849. Syntypes: Suriname: H. A. H. Kegel 191 (GOET), 457 (GOET)

Delima dasyphylla Miq. Stirp. Surin. Sel. 107. 1850. Nom. illegit. Type: Suriname: Hostmann & Kappler 701a (holotype U, isotypes P, S, W)
Delima dasyphylla Miq. fma. angustifolia Stirp. Surin. Sel. 108. 1850. Nom. illegit. Type: Suriname: Hostmann & Kappler 707 (holotype U, isotypes P, W)

Woody liana, young branches glabrous, often with grayish-white exfoliating bark. Leaves smooth on both sides, villous abaxially along veins, lanceolate to elliptic, 6-24 × 3-10 cm, apex acuminate, base attenuate, margin coarsely dentate, non-punctate; leaf blade with 11-18 pairs of lateral veins, ending at margins in a sharp tooth; petioles 1-2.5 cm long, villous, very narrowly alate. Inflorescences axillary, fasciculate with 5-30 pedunculate flowers; peduncles 1-2.5 cm long, sparsely strigose. Flowers 0.7-1.2 cm broad; sepals unequal, 2-5 mm long, elliptic to obovate, pubescent externally, glabrous inside; petals ca. 4 mm long, obovate, white, deciduous; stamens 25-40; carpels 1, glabrous; style ca. 1 mm long. Fruits 0.5-0.9 cm in diam., globose, red; seed 1, black, with a white aril.

KEY TO THE SUBSPECIES

Leaves lanceolate, cartaceus or subcoriaceus; surface glabrate or villous; lateral veins 11-22 . . . . 3a. *D. dentatus* subsp. *dentatus*
Leaves obovate or lanceolate-obovate; coriaceus; surface tomentulose; lateral veins 8-14 . . . . 3b. *D. dentatus* subsp. *esmeraldae*

3a. **Doliocarpus dentatus** (Aubl.) Standl. subsp. **dentatus.**

> *Tigarea dentata* Aubl. Hist. Pl. Guiane 2: 920. 1775. Type: French Guiana, Aublet, (BM).

Distribution: Central America, West Indies, Venezuela, Colombia, Ecuador, Peru, Bolivia, the Guianas, Brazil (over the Amazon, Northeast, Brazil Central and Southeast) and Paraguay. Wet forests and semideciduous forest, riverside, savannas and disturbed areas. 120 collections studied, 24 collections from the Guianas (GU: 8; SU: 12; FG: 4).

Selected specimens: Guyana: Demerara R. 06° 50' N; 58° 25' W., Boom 7187 (NY); Takutu R., Kanuku Mountains, Smith 3273 (F, G, NY, S, US, W). Suriname: Marowijne R., Arminafalls, Lanjouw 532 (U); Paulus Cr., Mennega 232 (NY, U). French Guiana: Maroni, Mélinon s.n. (P, R).

Phenology: Flowers reported March and April, fruits from August to September.

Vernacular names: Suriname: dia tetee.

3b. **Doliocarpus dentatus** (Aubl.) Standl. subsp. **esmeraldae** (Steyerm.) Kubitzki, Mitt. Bot. Staatssamml. München. 9: 57. 1971. –*Doliocarpus esmeraldae* Steyerm., Fieldiana, Bot. 28: 366. 1952. Type: Venezuela, Amazonas, Esmeraldas, Steyermark 57879 (F, IT: NY, VEN).

Distribution: Venezuela, Suriname, French Guiana, Brazil (Goiás) and Bolivia; wet forests and borders of forest and sandy soils savanna, from 100 up to 1400 m alt.; 20 collections studied, 4 collections from the Guianas (SU: 1; FG: 3).

Selected specimens: Suriname: Jodensavanne, Mapane Kreek area, Lindeman 4131 (NY, U). French Guiana: Cayenne, Sinnamary R., above Petit Saut, 05° 00' N; 53° 01' W, 87 m, Mori *et al.* 23434 (NY, PORT).

Phenology: Flowering in October.

4. **Doliocarpus gracilis** Kubitzki, Mitt. Bot. München 9: 716. 1976. Type: Brazil, Pará, without collector, herbarium nr. MG 10155 (MG, M).

Liana, branches and branchlets sulcate, laxe puberulent, bark flaking off. Leaves subcoriaceous, obovate, rarely elliptic, 2-7 × 2.5-5 cm, apex obtuse or acute, base cuneate-anguste, margin subrevolute, sinuate-dentate, mostly on the upper half; leaf blade glabrous, punctuate abaxially, with 7-10 pairs of lateral veins, convergent towards margin and linking 1.5-2 mm close to the margin; petiole 2-4 mm long, subwinged, minutely pubescent. Inflorescence axillary, fasciculate, 1-flowered; peduncles thin, 0.8-2 cm long, glabrous; sepals 3, oblong, 2-4 mm long, sparsely appressed, pilose to glabrous outside, glabrous inside; petals not seen; stamens ca. 20, anthers ca. 5 mm long; ovary ca. 3 mm long, glabrous. Fruit subglobose, 7-8 mm long.

Distribution: Suriname, French Guiana and Brazil (Amapá, Maranhão, Pará). Lowland wet forest. 8 collections studied, 4 collections from the Guianas. (SU: 1; FG: 3).

Selected specimens: Suriname: Nassau, Lanjouw & Lindeman 2821 (NY). French Guiana: St. Petit Japigny, 03° 30' N; 52° 53' W, Oldeman 1030 (CAY, M); Fl. Oyapock, Oldeman 1752 (CAY).

Phenology: Flowers reported in October, fruits in May.

5. **Doliocarpus guianensis** (Aubl.) Gilg, in Engler & Prantl, Nat. Pflanzenfam. 3(6): 114. 1893. – *Soramia guianensis* Aubl., Hist. Pl. Guiane 1: 552. 1775. Type: French Guiana, Aublet s.n. (BM).

Liana, branches and branchlets terete, laxe pilose, glabrescent when mature. Leaves coriaceous, glabrous on both sides, obovate or obovate-oblong, 6-21 × 2.8-9.5 cm, apex rounded or obtuse-apiculate, base cuneate, margin subrevolute, entire or subsinuate, mostly in the upper half; leaf blade with 7-10 lateral veins, anastomosing very near the margin or running straight towards the margin; midrib and lateral veins slightly impressed adaxially, prominent abaxially, tertiary veins prominent on both sides; petioles 5-10 mm long, narrowly winged, canaliculate, minutely pilose. Inflorescences axillary, racemoses (2-6 flowered), 1.5-4.5 cm long, minutely pilose to glabrescent when mature, bracteoles ovate-lanceolate, ca. 2 mm long, appressed sericeous at both sides; sepals 5, pilosule outside, glabrous inside, 3-5 mm long; petals 3-5, white, crenulate at the apex, soon deciduous; stamens ca. 60; filaments filiform, dilated towards the top; anthers ovate; carpel 1, globose, glabrous; rarely sparsely pilose, style curved; stigma peltate. Fruit ovoid-globose, 9-12 mm long, 10 mm wide; seeds 1-2, seeds reniform, ca. 9 mm long.

Distribution: Panama, Venezuela, the Guianas and Brazil (Amazonas). Wet low forest, dominated by *Mora spp*. 200-350 m. alt.. 35 collections studied, 18 collections from the Guianas. (GU: 2; SU: 3; FG: 13).

Selected specimens: Guyana: Potaro R., Cowan & Soderstron 2111 (F, NY, US). Suriname: Maratakka R., Maas 10819 (U). French Guiana: Ile de Cayenne, Kubitzki 71-124 (M); St. Jean du Maroni, Benoist 951 (U); Bassin du Counana, 04° 35' N; 52° 35' W, Granville 9112 (B, CAY, HBG, NY, P. PORT, U, US, USM)

Phenology: Flowers reported from April until June, fruits from December to January.

Uses: The water extracted from the stems of this species is potable and used in popular medicine as a diuretic, for conjunctivitis, and for kidney diseases. In addition, the seeds and aril edible when ripe in Guyana.

6. **Doliocarpus macrocarpus** Mart. ex Eichl. in Mart., Fl. Bras. 13(1): 77. 1863. Type: Brazil, Pará, Martius s.n. (M).

Doliocarpus virgatus Sagot, Ann. Sci. Nat., Bot. sér. 6, 10: 381. 1880. Syntypes: French Guiana: Perrottet 71 (G, P), Melinon s/n (P).

Liana, branches and branchlets grey, covered by long appressed hairs. Leaves subcoriaceous, obovate, obovate-oblong or obovate elliptical, 6-20 × 3-11 cm, apex rounded, sometimes abruptly acuminate, base cuneate, margin often revolute, entire, or sinuate-denticulate mostly in the upper half; leaf blade covered with appressed hairs adaxially, tomentose abaxially, pilose on the midrib and with some scattered hairs, with 10-

14(-18) lateral veins running straight towards the margin, subimpressed adaxially, prominent abaxially; petioles 0.7-2 cm long, subwinged, strigulose. Inflorescence racemose, 0.5-3 mm long, produced from short shoots (brachyblastus), axillary, peduncle tomentose; bracts suborbicular, appressed pilose outside, glabrous inside; sepals 4-5, 2-5 mm long, obovate or elliptic, sparsely pilose outside, glabrous inside; petals 4-5, soon deciduous, outside sparsely villose, inside glabrous; suborbicular or ovate-oblong, irregularly sinuate-angulate; stamens ca. 40, ca. 3 mm long, dilated towards the apex, anthers oblong; carpel 1, 2-ovuled, subglobose, densely villose; style ca. 2 mm long; stigma peltate. Fruits globose, ca. 8 mm in diameter, sparsely appressed pilose; seeds 2, reniform-globose, enclosed by a white, fleshy aril.

Distribution: Colombia, Venezuela, Peru, the Guianas and Brazil (Acre, Amazonas, Pará). Wet low forest and forest borders with sandy soils savanna. 15 collections studied, 6 collections from the Guianas. (GU: 1; SU: 2; FG: 3).

Selected specimens: Guyana: Bootooba, Demerara River. Persaud 28 (F, NY); Suriname: Brownsberg, BW 6427 (K, U); Suriname: Brokopondo, Village Brokopondo, van Donselaar 3864 (NY, U). French Guiana: Maroni, Mélinon 1864 (P).

Phenology: Flowers reported in August, fruits from March to April.

7. **Doliocarpus major** Gmel. Syst. Nat. 1: 805, 1791. Type: Suriname, Rolander s.n. (SBT).

In the Guianas only: subsp. **major**

Delima guianensis Rich. ex DC., Syst. 1: 408. 1817 (1818). Type: French Guiana: Richard s/n (P).
Doliocarpus spinulifer Miq. Linnaea 18: 266. 1844. Syntypes: Suriname: Focke 364 (U), Focke 814 (U).

Liana, branches and brachlets angulate, puberulent, glabrescent when mature, bark flaking off. Leaves charthaceous or subcoriaceous, lanceolate, lanceolate-elliptic or lanceolate-ovate, 4.5-16 (-20) × 2-7 cm, apex acuminate, base rounded, margin serrate or sinuate-mucronate; leaf blade sparsely villose adaxially, punctate and sparsely strigose abaxially, appressed pubescent along the veins, with 6-9 pairs of lateral veins, these extending and connected near the margin or running straight towards the margin; petioles 0.2-1.5 cm long, canaliculated, appressed pubescent, subwinged. Inflorescences axillary, composed of fascicles of 2-6(16) flowers; peduncles 0.2-2.5 cm long, laxe pubescent. Flowers 1-1.5 cm broad; sepals 5, unequal, obovate or elliptic, outer ones 2-3 mm long, inner ones 4-7 mm long, pubescent externally, appressed pilose inside; petals ca. 1.5 mm long, white, unguiculate at the base; stamens 70-80; carpel 1, villose; style 0.6-0.8 cm long, glabrous. Fruit 1-1.3 cm in diam., subglobose, pubescent, with hairs ca. 0.5 mm long, red; seeds 1-2, black, covered with a white aril.

Distribution: Nicaragua, Costa Rica, Panama, Colombia, Venezuela, Peru, Bolivia, the Guianas and Brazil (over the Amazon, Brazil Central and Northeast); wet evergreen forests. 200-400 m alt.. 70 collections studied, 33 collections from the Guianas. (GU: 3; SU: 21; FG: 9).

Selected specimens: Guyana: Cuyuni R., Sandwith 686 (K, NY, U). Suriname: Saramaca R., Maguire 23863 (F, NY, U, US); Paramaribo, Kegel 122 (GOET). French Guiana, Maroni, Sagot 1101 (G, K, P, S).

Phenology: Flowers reported in April and May, fruits from July until October

Vernacular name: Suriname: dija tetee, watra houtet. French Guiana: makoinde-tetei, taki-taki.

8.   **Doliocarpus multiflorus** Standl., J. Wash. Acad. Sci. 15: 285. 1925.
     Type: Panama. Colón, Canal Zone, between France field and Catival, Standley 30285 (K, US!).

Liana, branches and branchlets densely pilose, glabrescent when mature, bark red-brown, flaking off. Leaves subcoriaceous, obovate-lanceolate, 10-20 × 4-10 cm, apex obtuse or rounded, base cuneate, margins mucronate-serrate mostly in the upper half; leaf blade glabrous on both sides, apressed pilose along the midrib and secondary veins on abaxial surface, with 9-11 lateral veins, the upper ones anastomosing very near the margin or running straight towards the margin, midrib and lateral veins slightly impressed adaxially, prominent abaxially; petioles 1-2 cm long, pilose, subwinged. Inflorescences racemoses (2-4 flowered), 1.5-4 cm long, laxe pilose to glabrescent, bracteoles ovate-lanceolate, ca. 2 mm long; sepals 4-5, 2-4 mm long, obovate or obovate-oblong, laxe pilose outside, glabrous inside; petals 3-5, white; stamens ca. 40; carpels 1, villose; style short, curved; stigma peltate. Fruit ovoid-globose, 10-15 mm long, red, sparsely apressed pilose; seeds 1-2, black, reniform, covered with a white aril.

Distribution: Central America, Venezuela, Suriname, Colombia, Ecuador, Peru, Bolivia and Brazil (Maranhao, Pará, Roraima); wet low forest, dominated by *Mora spp*. 200-350 m. alt.. 35 collections studied, 1 from the Guianas. (SU: 1).

Selected specimen: Suriname: 20 km ESE of Brownsweg, van Donselaar 2045 (BBS, U).

9. **Doliocarpus paraensis** Sleumer, Repert. Spec. Nov. Regni Veg, 39: 45. 1935. Type: Brazil, Pará, Sampaio 5154 (B).
   Doliocarpus surinamensis Lanj., Rec. Trav. Bot. Néerl. 37: 290. 1940. Type: Suriname, BW 12 (U).

Liana, branches and branchlets terete, glabrous, bark grey, flaking off. Leaves coriaceous, obovate-oblong, obovate-lanceolate, 8-18 × 3-7.5 cm, apex shortly acuminate, base cuneate, margin entire, revolute; leaf blade glabrous on both sides, punctate abaxially, with 8-11 lateral veins, midrib and lateral veins prominent on both sides, especially abaxially, conspicuously arcuate, anastomosing at about 4 mm from the margin; tertiary reticulation prominent at both sides; petioles 0.5-2 cm long, glabrous, narrowly winged. Inflorescences axillary, few flowered (2-3), pedicels ca. 5 mm long; sepals 4-5, 5-6 mm long, oblong, sparsely pilose to glabrescent inside, glabrous outside, margin ciliate, petals ca. 1 cm long, cuneate-spathulate, white; stamens 40-50; filaments 4-5 mm long, glabrous, slightly dilated towards the apex; anthers ca. 5 mm long, oblong; carpel 1, densely pilose, trichomes ca. 4 mm long. Fruit globose, 2-3 cm diameter, sparsely pilose to glabrescent, 1-seeded; seed black, reniform, ca. 1.5 cm long, enclosed by a white aril.

Distribution: The Guianas and Brazil (Pará). Wet forest, 200-400 m alt.; 20 collections studied of which 14 collections from the Guianas. (GU: 1; SU: 12; FG: 1).

Selected specimens: Guyana: Essequibo R., Onoro Creek, Smith 2756 (F, K, NY, P, S, US). Suriname: Brokopondo, van Doselaar 3486 (U); Nassau Mts., Laujouw & Lindeman 2200 (U); Lucie R., Hulk 404 (U). French Guiana: Camopi, Embouchure du Yaroupi, affluent de l'Oyapock, Oldeman 3427 (CAY, US).

Phenology: Flowers reported in December.

Vernacular name: Suriname: mabijara.

10. **Doliocarpus sagotianus** Kubitzki, Mitt. Bot. Staatssamml. München 9: 43. 1971. Type: French Guiana, Sagot 1155 (holotype P, isotypes BR, GH, K, NY, S, W).

Liana (?), branches and branchlets terete, glabrous. Leaves rigid-coriaceous, obovate-lanceolate, 6-12 × 2.5-5.5 cm, apex small rotundate or acute, base acute or subrotundate, margin entire, subrevolute; leaf blade glabrous adaxially, sparsely pilose abaxially, with 8-11 pairs of lateral veins, impressed adaxially, prominent abaxially, running straight towards the margin; petioles 0.6-1 cm long, sparsely pilose to glabrous, winged. Inflorescences racemose, 2-3 flowered, 3-8 mm long; peduncles and pedicels (ca. 2.5 mm long) erect pilose; sepals 6, obovate-elliptic, 2.5-5 mm long, sparsely pilose outside, glabrous inside; petals not seen; stamens ca. 50; carpel 1, glabrous. Fruit not known.

Distribution: Endemic, only known by the type and three collections.

Selected specimens: French Guiana: Sagot 1155 (BR, GH, NY, K, P, S, W).

11. **Doliocarpus savannarum** Sandw., J. Arnold Arbor. 24: 218. 1943. Type: Guayana: Kaieteur Savanna, Jenman 1038 (K).

Doliocarpus ptariensis Steyerm., Fieldiana, Bot. 28: 367. 1952. Type: Venezuela, Bolivar, Ptari-Tepui. Steyermark 59972 (holotype F, isotypes NY, VEN).

Erect or procumbent shrub to woody liana, branches and branchlets strigose, glabrous when mature. Leaves coriaceous, obovate, lanceolate-obovate, 7-20 (-28) × 3-10 cm, apex obtuse or acute, acumen ca. 0.5 mm long, base acute or obtuse, margin subrevolute, entire or denticulate; leaf blade glabrous adaxially, appressed pilose on midrid and lateral veins abaxially, with 9-14 (-16) pairs lateral veins, running straight towards the margin; petioles 0.8-2 cm, appressed pilose, subwinged. Inflorescences 1-6 mm long, produced from long shoots (dolichoblastus), 3-7-flowered, peduncles pubescent, bracteoles ovate-lanceolate; sepals 5, 2-4.5 mm long, elliptic, sparsely pubescent outside, glabrous inside; petals soon deciduos; stamens 40-50; carpel 1, densely pilose; style ca. 2 mm long, pilose. Fruit ovoid globose, sparsely villose, 0.8-1 cm diameter; seed black, covered by a white aril.

Distribution: Colombia, Venezuela, Guyana, Brazil (Amazonas, Roraima) and Bolivia; white sand soils (Roraima formations), wet savanna, 150-1600 m alt.. 25 collections studied, 13 collections from Guyana.

Selected specimens: Guyana: Kaiteuer Savanna, 400 m alt., Sandwith 1377 (NY, K); Mazaruni Flusses, Maguire 32651 (NY).

Phenology: Flowers reported from July to October, fruits from May to June.

Note: This species is morphologically variable. Specimens from low altitudes (100-600 m) have large and glabrescent leaves, while specimens from high altitudes (700-1600 m) have relatively smaller and pubescent leaves.

12. **Doliocarpus spraguei** Cheesm., Fl. Trinidad & Tobago 1: 473. 1947. nom nov.; Mitt. Bot. Staatssamml. München 9: 41. 1971. – *Calinea scandens* Aubl., Hist. Pl. Guiane 1: 556. 1775. Type: French Guiana, Aublet s.n (BM, photo NY).

Small shrub or woody liana; branches terete or subterete, glabrous. Leaves rigid-coriaceous, lanceolate, obovate-lanceolate, conduplicate (herbarium specimens), 0.5-12 × 2-6 cm, apex acuminate, base acute, margin entire, subrepanded; leaf blade glabrous on both sides, with 8-10 lateral veins, curved, running straight towards the margin or anastomosing at about 0.2 mm from the margin, midrib slightly impressed adaxially, prominent abaxially, tertiary reticulation distinctly prominent on both sides; petioles 0.5-1 cm long, canaliculated above. Inflorescences fascicles racemose, peduncles 0.5-1 cm long, sparsely pilose; bracteoles lanceolate-triangular; pedicels 2-4 mm long; flower ca. 5 mm in diameter; sepals 5, unequal, obovate-elliptic, 2-3.5 mm long, persistent, laxe pilose outside, glabrescent inside; petals 2-3, suborbicular, 3 mm long, soon deciduous; stamens 40-60; filaments filiform, slender, dilated towards the apex; anthers subglobose; carpel 1, subglobose, glabrous. Fruit 6-7 mm diameter, subglobose; seeds 1-2, ca. 4 mm long, black, subreniform, enclosed by a white aril.

Distribution: Venezuela, Trinidad and Tobago, the Guianas and Brazil (Amazonas, Goiás, Pará); white sand savanna and sclerophyllous forests, 200-400 m alt.. 80 collections studied, 45 collections from the Guianas. (GU: 18; SU: 20; FG: 7).

Selected specimens: Guyana: Berbice-Corentijne Region, Pipoly *et al.* 11565 (US); Demerara-Mahaica Region, Hahn *et al.* 3933 (US); Potaro-Siparuni Region, Gillespie *et al.* 1300 (US). Suriname: Paramaribo, Kramer & Hekking 2856 (U); Tafelberg, Maguire & Fanshawe 24284 (M, NY, US); Brownsberg, van Donselaar 2780 (U). French Guiana: Cayenne, Cremers 9426 (CAY, P); Ouaqui, Bafog 7826 (U).

4. **NEODILLENIA** Aymard, Harvard Pap. Bot. 10: 123. 1996. (1997).
Type: *Neodillenia coussapoana* Aymard

Lianas; stems with vascular bundles disposed in bands or concentric rings, separated by abundant parenchyma. Leaves smooth, elliptic, broadly ovate, orbicular, obovate, rounded at the apex, rounded or obtuse at the base, with 6–18 pairs of primary lateral veins, connected near the margin or running straight towards the margin; stipules caducous; tertiary venation parallel, margins entire, subsinuate or dentate; petiole not winged or subwinged, thick at the base. Inflorescence ramiflorous, axillary, fasciculate, few-flowered, racemose or reduced to a solitary flower; bracts orbicular. Flowers bisexual, pedicellate or sessile; sepals 3–6, unequal, imbricate, orbicular, wide ovate or obovate, covering the young fruit; petals fugacious, not seen; stamens 100–300 or more, forming a rim around the carpels; filaments free; anthers basifixed; thecae linear, glabrous, opening by 2 longitudinal slits. Carpels 1–5, free, connivent; styles 1–5; stigma peltate, glabrous; ovules 1 or 2 by carpel, campylotropous or orthotropous, ascending from the basal placenta, the raphe ventral. Fruits capsular. Seeds 1 or 2, black, reniform; aril entire, red.

Distribution: *Neodillenia* is a genus with six species (three still undescribed) found at low to middle elevations in the Amazonian basin of Colombia, Venezuela, Ecuador, Peru, Brazil, and French Guiana.

Note: The genus *Neodillenia* had been a notorius matter inside of the Dilleniaceae. Kibutzki (2004) mereged it into *Doliocarpus* without any explanation. Subsequently, Horn (2006) categorized the *Neodillenia* as a genus *dubium*, and placed it in subfamily Doliocarpoideae based mainly on its successive cambia, peltate stigma, and ovular details, arguing in addition that the stamens were free, and the ovules campylotropous. Later, in a phylogenetic analysis of Dilleniaceae, Horn (2009) provisionally placed *Neodillenia* in Doliocarpoideae, although no species of the genus were represented in this study; the author, perhaps proposed his assessment based on an array of Doliocarpoideae character states that justify the placement of *Neodillenia* in this subfamily. the most recent phylogenetic study of the neotropical Dilleniaceae (Fraga, 2012), did include *N. coussapoana* Aymard. The results clearly indicate that *Neodillenia* forms a clade with *Davilla* that is basal to *Doliocarpus* and *Curatella*. These latest results based on DNA sequences indeed confirm that *Neodillenia* belong in Doliocarpoideae. It resembles *Doliocarpus* in the vascular bundles disposed in bands or concentric rings separated by abundant parenchyma, the connective linear (present in only two species of *Doliocarpus*: *D. grandiflorus* Eichl. in Mart. and *D. magnificus* Sleumer), inflorescences ramiflorous, and one carpel (only in *Neodillenia venezuelana* Aymard). Nevertheless, *Neodillenia* definitely does not belong in *Doliocarpus* because of its fasciculate inflorescences reduced to a single flower, larger flower buds like *Dillenia* (1-4 cm long), sepals unequal, 1.5-4 cm long, covering the fruit; stamens 100-300 or more, forming a rim around the carpels; anthers 4-6 mm long, connective always linear, carpels 1-5, and aril red. In contrast, *Doliocarpus* always has ramiflorous inflorescences, fasciculate or racemose, not reduced to a single flower, flower buds 2-8 mm long, sepals 0.3-1.8 cm long, never covering the fruit; stamens 25-80, not forming a rim around the carpels; anthers 0.5-3 mm long, connective broadened and sometimes thickened; carpel always one, and aril white.

LITERATURE

Aymard, G. 1997. Dilleniaceae Novae Neotropicae IX. Neodillenia a new genus from Amazon basin. Harvard Papers in Botany 10:121-131.

Fraga, N. C. 2012. Filogenia e revisão taxonômica de Davilla Vand. (Dilleniaceae). Ph.D Thesis, Universidade Federal de Minas Gerais, Instituto de Ciências Biológicas, Departamento de Botânica. Belo Horizonte, Brasil, 423 pp.

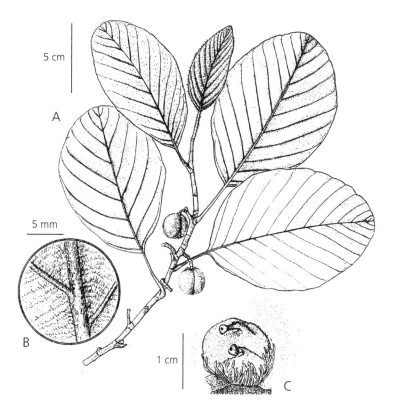

Fig. 4. *Neodillenia coussapoana* Aymard: A, Inflorescence branch; B, Lower leaf surface; C, Flower with the sepals removed, leaving a horizontal scar. From Aymard C., G.A. 1997. Dilleniaceae novae Neotropicae IX, Neodillenia, a new genus from the Amazon basin. Harvard Pap. Bot. 10, 121-131.

Neodillenia sp A.

Liana; branches and branchlets completely covered with golden-yellow trichomes, sparsely pilose when mature. Leaves coriaceous, elliptic, 20-30 cm long, 10-16 cm wide, rounded at the apex, obtuse at the base, margins entire to subsinuate, mostly in the upper half, soft pilose on the upper surface, with erect yellow trichomes, the midrib and secondary nerves canaliculated and covered by an appressed pubescence, dense pilose on the lower surface, the midrib and secondary nerves densely golden-yellow appressed pubescent, lateral nerves 10--13, convergent close to the margin; tertiary venation parallel; petiole 3-4.5 cm long, 3-5 mm wide, unwinged, canaliculate, thick at the base, densely golden-yellow appressed pubescent, sparse when mature. Inflorescence ramiflorous, axillary, fasciculate, or reduced to a solitary; peduncle densely yellow-tomentose, 2-3.5 cm long; bracts not seen. Sepals 5, unequal, imbricate, coriaceous, densely yellow-appressed pubescent externally, glabrous internally, the outer 2, orbicular, 2--3 cm long, 2--2.5 cm broad, the inner 3, ovate, ca. 2 long, 1.5-2 cm broad; petals not seen; stamens 300 or more, ca. 5 mm long, forming a rim around the carpels; filaments ca. 2 mm long, free, glabrous; anthers ca. 3 mm long, oblong, glabrous, opening by two longitudinal slits. Carpels 2, not striate, globose, sparsely apressed pilose, styles and stigma not seen; ovules 1 by carpel. Fruit a capsule covering by the sepals; seed black, shiny, reniform, ca. 1 cm long, completely covered by a red aril.

Distribution: Endemic, only known by two collections from French Guiana lowlands

Selected specimens: French Guiana: Fleuve Oyapock, Trois Sauts, Wayampi Zidock Ville, 02° 15' N; 52° 53' W, 0 m, C. Sastre 4681 (CAY, NY, P, U). Trois Sauts, H. Jacquemin 1875 (CAY, P).

Vernacular names: Tameyu (Wayampi).

5. **PINZONA** Mart. & Zucc., Abh. Math. -Phys. Cl. Königl. Bayer. Akad. Wiss. 1: 371. 1832.
   Type: *Pinzona coriacea* Mart. & Zucc.

Thick-stemmed lianas; stems with vascular bundles disposed in bands or concentric rings and separated by abundant parenchyma. Stipules absent. Leaves alternate, elliptic or obovate; petioles winged; margins entire, rarely serrate. Inflorescences racemes or paniculate arrangements of racemes, axillary, pilose; bracts lanceolate to ovate. Flowers bisexual; sepals 3 or 4, subequal, glabrous inside, margins pilose; petals 2-3, white, obovate, subemarginate; stamens 25-40; carpels 2, globose, glabrous, connate ventrally from base to apex of ovary; ovules 2 in each carpel, anatropous; styles 2, terminal; stigma peltate. Fruits paired capsules; seeds 1-2, obovate, arillate, aril orange.

Distribution: Central America, Antilles, Colombia, Venezuela, Guyana, Suriname, French Guiana, Brazil (Amazonas, Pará, Roraima), Ecuador to northern Peru.

1. **Pinzona coriacea** Mart. & Zucc., Abh. Math.-Phys. Cl. Königl. Bayer. Akad. Wiss. 1: 371. 1832. Type: Brazil, Amazonas: Brasilia aequatoriali ad flumen, Martius s.n (M, not seen).   – Fig. 5

Doliocarpus nicaraguensis Standl., Field. Mus. 4:233. 1929. Type: Nicaragua, region of Bragman's Bluff, Englesing 277 (holotype, F, isotype K).
Doliocarpus belizensis Lundell, Field & Lab. 13: 6. 1945. Type: Belize, Toledo District, between Rancho Chico and Cockcomt, Monkey R., Gentle 4389 (holotype SMU, isotypes NY, US).

Thick-stemmed liana climbing up to 25 m or higher; stems diameter more than 20 cm circumference with vascular bundles disposed in bands or concentric rings and separated by abundant parenchyma. Branches and branchlets angulate, sparsely apressed pubescent, glabrescent when mature, bark reddish, flaking off in large chunks when mature. Leaves coriaceous, elliptic, obovate or obovate to ovate, rarely lanceolate, 6-16 × 5-12 cm, apex rounded to obtuse, base rounded to obtuse or cuneate, often shortly apiculate, margins subrevolute, entire, sinuate, mostly in the upper half; leaf blade glabrous on both sides, except along the midrib and secondary veins, where is sparsely apressed pubescent, not papillate, with 7-9 lateral veins, convergent and linking close to the margin; petioles angulate, winged, 1-2.5 cm long, apressed pubescent, glabrecent when mature. Inflorescences racemes or paniculate arrangements of racemes, axillary, 3-7 cm long, rachis strigose-pubescent, glabrescent when mature; bracts lanceolate to ovate, 1-3 mm long, apressed pubescent externally, glabrescent internally. Flowers pedicellate, pedicels 2-8 mm long, strigose pubescent, glabrescent when mature; sepals 3 or 4, subequal, ovate to broadly ovate-elliptic, 3-5 mm long, sparsely pilose externally, glabrous internally, margins pilose; petals 3, white, deciduous, oblong-obovate, subemarginate, 4-5 mm long, glabrous externally, glabrous internally except along the midrib, where is sparsely apressed pubescent; stamens 25–40, filaments slender, glabrous, 2.5-25 mm long, anthers ca. 5 mm long, glabrous; carpels 2, globose, glabrous, connate ventrally; ovules 2, anatropous; styles 2, terminal, 2-2.5 mm long, persistent at fruit; stigma peltate. Fruits glabrous, bilobed capsules, 3-5 × 5-7 mm, splitting open at right angles to juncture between carpels; seeds 2, one per carpel, black, shiny.

30

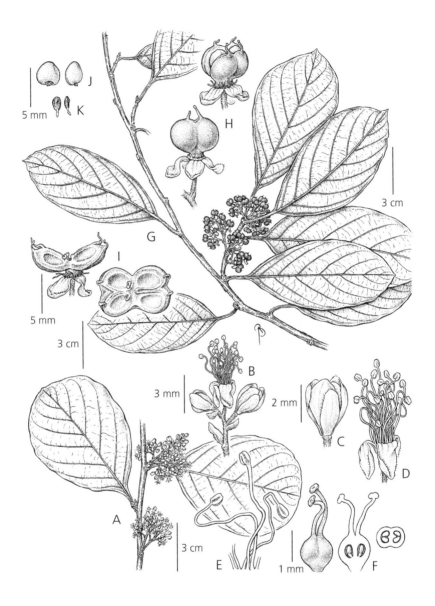

Fig. 5. *Pinzona coriaceae* Mart. & Zucc.: A, Flowering branch; B, Detail of inflorescence; C, Bud; D, Flower; E, Stamens; F, Gynoecium, entire, longitudinal section, and cross section; G, Fruiting branch; H, Fruit, closed and open; I, Open fruit, without seeds, side and top view; J, Seed, front and side views; K, Embryo, front and side views. From Mori, S.A. *et al.* 2003. Vascular plants of central French Guiana. Mem. NYBG Vol. 76(2).

Distribution: Central America, Antilles, Colombia, Venezuela, Guyana, Suriname, French Guiana, Brazil (Amazon Basin), Ecuador to northern Peru; wet low forests 200-400 m alt.. 60 collections studied, 11 collections from the Guianas (GU: 6; SU: 2; FG: 3).

Selected specimens: Guyana: Potaro R., Maguire & Fanshawe 23066 (M, NY, US). Suriname: Corantijn R. and Kaboerie, Lindeman 5198 (U). French Guiana: L'Inini, Mont Atachi Bacca, 10 km SE de Gobaya Soula, 53° 55' N; 3° 33" W, de Granville *et al.* 10789 (CAY, NY, PORT, US).

6. TETRACERA L., Sp. Pl. 1: 533. 1753.

Type: *T. volubilis* L.

Subshrubs, slightly woody to woody lianas, stems more than 30 cm circumference. Stipules absent. Leaves simple, alternate, pinnately veined, scabrous with hairs simple or fasciculate. Plants androdioecious (neotropical species). Inflorescence racemes (thyrses, cincinnus) terminal or lateral, bracts lanceolate. Flowers actinomorphic, fragrant, polygamous; sepals 4-7(-12), orbiculate, persistent, imbricate, glabrous, strigose or sericeous; petals 3-5, obovate or oblong; stamens numerous, 50-200; filaments filiform; anthers short, expanded; carpels 1-5, free; style very short; stigma peltate; ovules anatropous. Fruit folicles coriaceous, disposed; seeds 1-4 per locule, black, arillate, aril laciniate, red or orange.

Distribution: Pantropical genus, with approximately 16 neotropical species, 13 in Africa and 15 in Australia-Asia. In the neotropics it occurs in Mexico, Central America, Antilles, Venezuela, the Guianas, Colombia, Ecuador, Peru, Bolivia, Brazil and Paraguay.

Notes: Kubitzki work (1970) is by far the noteworthy account of *Tetracera*; in his revision of American and African species, he used inflorescences, flowers, fruit, and trichomes characters to divide the genus into two sections (*Tetracera* and *Akara*). Moreover, the section *Tetracera* was divided into groups (*volubilis* and *willdenowiana*), categories that were confirmed by phytochemical data (Kubitzki, 1970).

LITERATURE

Aymard, G.A. and B.M. Boom. 2002. A new species of Tetracera (Dilleniaceae) from Guyana. Brittonia 54(4): 275-278.

Barrie, F.R. & C.A. Todzia. 1991. Proposal to conserve Tetracera volubilis L. and its type (Dilleniaceae). Taxon 40(4): 652-655.

Kubitzki, K. 1970. Die Gattung Tetracera (Dilleniaceae). Mitt. Bot. Staatssamml. München 8: 1-7218.

KEY TO THE SPECIES

1. Lateral veins of the leaves running straight towards the margin (craspedodromous); sepals sericeous internally......................
...............................6. *T. volubilis* subsp. *volubilis*
Lateral veins of the leaves convergent towards margin and linking close to the margin (brochidodromous); sepals internally glabrous, sometimes with sparse trichomes but never with a sericeous pubescence............ 2

2. Erect shrubs; leaves rigid-coriaceous, apex emarginate, base cuneate; rachis and pedicel with simple trichomes................... 3. *T. maguirei*
Liana to scandent shrubs; leaves chartaceous to coriaceous, apex acuminate or rotundate, base rounded, obtuse or cordate; rachis and pedicel with stellate trichomes 3

3. Leaves orbicular to obovate, or elliptic to broadly elliptic............. 4
Leaves lanceolate-elliptic, oblanceolate, elliptic, ovate, obovate (rarely), or oblong... 5

4. Leaves orbicular to obovate, 2-5 cm wide; sepals glabrous or sparsely stellate internally, the inner ones 3-4 mm long; petals 3-4 mm long; fruit longitudinally not sulcate ....5. *T. tigarea*
Leaves elliptic to broadly elliptic, 4-14 cm wide; sepals appressed stellate-pilose internally, the inner ones 6-8 mm long, petals ca. 7 mm long; fruit longitudinally sulcate ....4. *T. surinamensis*

5. Leaves chartaceous to subcoriaceous, glabrous on the lower surface; carpels sparsely ciliate .... 1. *T. asperula*
Leaves coriaceous, pubescent on the lower surface; carpels glabrous 6

6. Abaxial surface of leaves with simple trichomes; sepals glabrous internally
....7. *T. willdenowiana*
Abaxial surface of leaves densely fasciculate-tomentose; sepals appressed pilose internally ....2. *T. costata*

1. **Tetracera asperula** Miq., Linnaea 19: 133. 1847. Type: Suriname: Paramaribo Kappler 1703 (BR, G, S). – Fig. 6

Tetracera grandiflora Eichler Fl. Bras. 13(1): 92, t. 22, f. 2. 1863. Suriname: Jodensavannah, Wullschlägel 1672 (BR).

Liana to scandent shrub, young branches dark, sparsely appressed setose to glabrescent when mature, older branches turning grayish, bark fissured and cracked. Leaves coriaceous, elliptic (rarely ovate), (4-)7-12(-16) × 4-7.4 cm, apex acute rarely widely acuminate or rotundate, base acute sometimes rotundate, margin simple, entire, sub-revolute; leaf blade glabrous adaxially, very slightly scabrous abaxially, with (6-)7-9 pairs of lateral veins, venation brochidodromous; midrib glabrescent or nearly so, midrib and lateral veins not impressed adaxially, slightly prominent

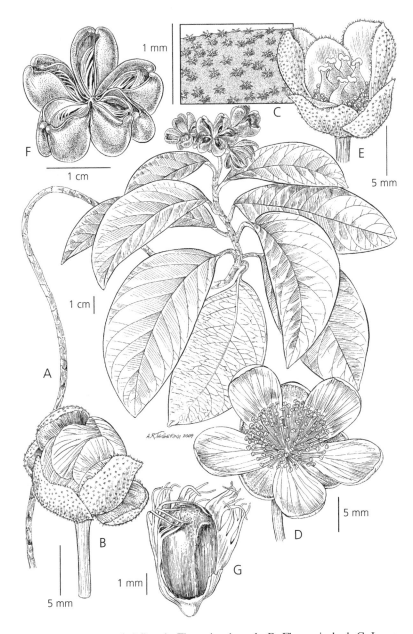

Fig. 6. *Tetracera asperula* Miq.: A, Flowering branch; B, Flower in bud; C, Lower leaf surface; D, Male flower, upper view; E, Female flower with immature fruit; F, Mature fruit; G, Seed with aril. Drawing by Alice Tangerini with permission of the Smithsonian Institution. (A, C, F-G, Christenson 1947; B, E, Cowan & Soderstrom 1802; D, Gillespie 2654).

abaxially; petioles 0.8-2.8 cm long, scabrous, glabrescent when mature, not winged. Inflorescence terminal or axillary, thyrse few to 20- flowered, 6-20 cm long, the rachis sparsely pubescent, with stellate trichomes. Flower androdioecious. Bisexual flowers: sepals unequal, broadly ovate, sparsely scabrous with a few stellate trichomes externally, sparsely appressed pubescent internally, the margin ciliated, outer sepals 3, 6-8 mm long, inner sepals 2, 8-10 mm long; petals 5, obovate, white, glabrous, 12-15 mm long. Staminate flowers: sepals unequal, broadly ovate, sparsely scabrous with a few stellate trichomes externally, glabrous internally, the margins ciliated, outer sepals 3, 4-6 mm long, inner sepals 2, 5-8 mm long; petals 5, obovate, white, glabrous, 12-15 mm long. Fruit 4-5 follicles, 1-1.4 cm long, glabrous or very sparsely ciliated, the style persistent, ca. 1 mm long; seed black, 2-3 per follicle, ca. 5 mm long; aril red, laciniate and almost completely covering the seed.

Distribution: Guyana, Suriname and French Guiana; 100 collections studied, all from the Guianas: (GU: 57; SU: 26; FG: 17).

Selected specimens: Guyana: Waranama Ranch, Intermed. savanna, Berbice R., Harrison & Persaud 1053 (BDG, K); Mt. Ayanganna, Pakaraima Mts., vic. Chinowieng, Maguire 40638 (NY). Suriname: Joden savanne, Mapanecreek area, E. of Suriname R., Kramer & Hekking 2466 (NY); Tibiti savanne on savanna near km 3.8, 3rd line, Lanjouw & Lindeman 1782 (NY, K). French Guiana: Commune de Iracoubo, Bassin de l'Iracoubo, W. Iracoubo, project de reserve, Cremers 9628 (US, CAY); Mana, Mélinon 90 (US, K, P, BM).

Vernacular names: Guyana: kabuduli (A), mibie (AR). Suriname: savanne-kautele

2. **Tetracera costata** Mart. ex Eichler, Fl. Bras. 13(1): 86. 1863. Type: Suriname: Paramaribo Wullschlägel 1342 (BR; isotypes: GOET, U).

Tetracera trinitensis Sprague & Williams, Fl. Trinidad & Tobago 1: 10. 1928. Type: Trinidad, Crueger s.n. (K).

Liana, young branches scabrous to densely appressed tomentose pubescence to sparsely appressed pubescent when mature, older branches grayish with numerous cracks and fissures, glabrescent. Leaves coriaceous, widely elliptic or oblong, 7-16 × 4-9 cm, apex bluntly acute to truncate, base petiolate, rotundate or broadly obtuse or acute (rarely subcordate), margin simple, entire, revolute; leaf blade scabrous adaxially, scabrous or tomentose abaxially, with 9-15(-19) pairs of lateral veins, venation brochidodromous; midrib and lateral veins impressed adaxially, appressed pubescent to glabrescent and prominent abaxially; petioles 1-3(- 4) cm long, winged, ca. 0.5-1 mm long, appressed pubescent or scabrous. Inflorescence terminal or axillary, thyrse many-flowered, 8-11 cm long, the rachis covered with dense appressed pubescent or stellate trichomes. Flowers androdioecious. Bisexual flowers: sepals unequal, ovate, scabrous externally sometimes with stellate trichomes, scabrous or appressed pubescence near apex, margins ciliated, outer sepals 3, 5-7 mm long, inner sepals 2, 6-8 mm long; petals 5, obovate, white, scabrous externally, short pubescence internally, ca. 5 mm long. Staminate flowers: sepals unequal to subequal, ovate, scabrous or appressed pubescence externally, grabrous or appressed pubescence internally, ciliate at margins, outer sepals 3, 2-3 mm long, inner sepals 2, 4-5 mm long; petals 4 obovate, white, glabrous, ca. 5 mm long. Fruit 2-4 follicles, 0.8-1.0 cm long, glabrous, the style persistent, ca. 3-4 mm long; seed black, 1 per follicle, ca. 4 mm long, aril red, laciniate and completely covering the seed.

KEY TO THE SUBSPECIES

1. Young branches densely appressed tomentose, sepals appressed pubescence
    . . . . 2a. *T. costata* subsp. *costata*
    Young branches scabrous, sepals scabrous, sometimes with stellate
    trichomes . . . . 2b. *T. costata* subsp. *rotundifolia*

2a. Tetracera costata Mart. ex Eichler subsp. **costata**

Young branches densely appressed, tomentose pubescence. Leaves tomentose abaxially; petioles sparsely appressed pubescence. Bisexual flowers: sepals more or less equal, imbricate, stellate, pubescent externally, scabrous internally; outer sepals 3 (rarely 4), ca. 7 mm long, ovate to orbicular; inner sepals 2, ca. 8 mm long, obovate; petals 3, whitish, obovate, ca. 11 mm long with pilose pubescent on both sides. Staminate flowers: sepals appressed tomentose pubescence on both sides.

2b. Tetracera costata Mart. ex Eichler subsp. **rotundifolia** (J. E. Smith) Kubitzki, Mitt. Bot. Staatssamml. München 8: 74. 1970. – *Tetracera rotundifolia* J. E. Smith., Cycl. 35(7). 1817. Lectotype: Guyana, Rudge s.n. (LINN).

Tetracera ovalifolia DC. Syst. Nat. 1: 400. 1817. Type: French Guiana: Patris s.n (G).

Young branches scabrous. Leaves scabrous, glabrescent when mature, glabrous or with sparsely stellate trichomes abaxially; midrib and lateral veins appressed pubescent to glabrous; petioles glabrous or nearly so.

Bisexual flowers: sepals more or less equal, imbricate, orbicular; outer sepals 2, 7-9 mm long; inner sepals 2, 9-10 mm long, glabrous (rarely sparsely pubescent) externally, tomentose interally, margin slightly ciliated, petals whitish. Staminate flowers: sepals subequal, obovate to ovate, sparsely scabrous with stellate trichomes externally, scabrous with short stiff trichomes internally, margins ciliated, outer sepals 3, 7-8 mm long, inner sepals 2, ca. 9 mm long.

Distribution: Evergreen lowland forest 200 - 400 m alt.. 13 collections studied, all from the Guianas: (GU: 5; SU: 1; FG: 7).

Selected specimens: Guyana: Mazaruni Station, Fanshawe F1179 (Record # 3915); Suddie, Essequibo, Stockdale 8791 (K). Suriname: no location, Hostmann 38709 (US type photo B). French Guiana: Ile Portal, Sagot s.n. (US); Cayenne, Brown s.n. (BM)

Vernacular names: Guyana: katuwai (AR)

3. **Tetracera maguirei** Aymard & Boom, Brittonia 54(4): 275. 2002. Type: Guyana, Mt. Ayanganna, Pakaraima Mts., vicinity of Chinowieng, 1000-1200 m alt., Maguire 40620 (NY!)

Small shrub, young branches sparsely appressed pubescent to glabrescent when mature. Leaves coriaceous, oblanceolate or rarely obovate, 1.5-14 × 1-4 cm, apex bluntly acute or rotundate, emarginate, base cuneate, margin simple, entire, strongly revolute; leaf blade glabrous on both sides, with 5-9 pairs of lateral veins, venation brochidodromus at margins; midrib sparsely appressed pubescent, glabrescent when mature, midrib and lateral veins slightly pubescent adaxially, prominent abaxially; petioles 2-4 mm long; winged, ca. 2-3 mm wide and glabrescent. Inflorescence terminal or axillary, thyrse few-flowered, 2-7 cm long, the rachis sparsely appressed pubescent, with simple trichomes. Flowers androdioecious. Bisexual flowers 5-merous; sepals unequal, imbricate, broadly ovate to orbicular, outer sepals 4-6 mm long, inner sepals 5-7 mm long, papillate and sparsely appressed pubescent externally, glabrous internally, sparsely appressed pubescent close to margin, margin ciliate; petals 5, imbricate, broadly ovate to orbicular, white, glabresent or sparsely pilose on both sides, 5-7 mm long. Staminate flowers 5-merous, sepals unequal, broadly ovate, papillate and sparsely appressed pubescent externally, glabrous internally, the margins ciliate, outer sepals 3, 4-6 mm long; inner sepals 2, 5-7 mm long; petals 5, obovate, white, glabrous, 10-15 mm long. Fruit 4-7 follicles, 9-16 mm long, glabrous or sparsely ciliate, the style persistent; seeds yellow, 1 per follicle, ca. 5 mm long; aril orange, laciniate and completely covering the seed.

Distribution: Endemic to Guyana; between 500-1200 m alt.; 3 collections studied, all from Guyana.

Specimens examined: Guyana: Mt. Ayanganna, Pakaraima Mountains, vicinity Chinowieng, Maguire 40620 (NY); Pakaraima mountains, Imbaimadai Cr., W of Imbaimadai, Pipoly & Alfred 7989 (US, NY); Karowrieng R., Maipuri Falls, Gillespie & Smart 2911 (US).

Vernacular names: Guyana: katuwai (AK)

4. **Tetracera surinamensis** Miq., Linnaea 18: 265. 1844. Type: Suriname: Cottica R., Focke 694 (U).

Tetracera ovalifolia de Candolle, Syst. Nat. 1: 400. 1817. French Guiana, Patris s.n. (G).

Liana, young branches scabrous to glabrescent when mature, older branches with numerous cracks and fissures. Leaves coriaceous, elliptic to obovate, 8.6-14 × 6-10 cm, apex bluntly acute or rotundate, emarginate, base slightly cordate, rotundate or truncate, margin simple, entire, sub-revolute; leaf blade glabrous adaxially, scabrous abaxially, with 8-11 pairs of lateral veins convergent towards margin and linking close to the margin (brochidodromous); midrib and lateral veins glabrous or with sparse appressed setose pubescent, elevated below, impressed above; petioles 2-2.4 cm long, winged, 1-2 mm wide, sparsely pubescent. Inflorescence terminal or axillary, thyrse few-18 flowered, 10-20 cm long, the rachis sparsely appressed pubescent mixed with stellate trichomes. Flowers androdioecious. Bisexual flowers: sepals unequal, imbricate, ovate to obovate; outer sepals 3, (4-)5-6 mm long, inner sepals 2, 7-8 mm long, scabrous externally, glabrous internally, the margin ciliated; petals 4, broadly ovate, white, glabresent on both sides, 5-6 mm long. Staminate flowers: sepals unequal, widely ovate, appressed pubescent externally and internally, the margin ciliated, outer sepals 3, 3-4 mm long, inter sepals 2, 6-8 mm long; petals 5, obovate, white, glabrous, 9-10 mm long. Fruit 3-4 follicles, 8-10 cm long, grooved, glabrous, style persistent, ca. 0.8 mm long; seeds black, 1 per follicle, ca. 0.5 mm long; aril laciniate and covering ca. half of the seed.

Distribution: Evergreen lowland forest 50-400 m. 11 collections studied, all from the Guianas: (GU: 7; FG: 4).

Selected specimens: Guyana: Upper Mazaruni R., de la Cruz 2371 (US, NY); Cow savanna along S bank of Canje R., W of Digitima Cr., Pipoly & Gharbarran 9480 (US). French Guiana: Karouany, Sagot 15 (K, P).

Vernacular names: Guyana: kabaduli (Arawak), katuwai (AK)

5. **Tetracera tigarea** de Candolle, Syst. Nat. 1: 403. 1818 (1817). – *Tetracera aspera* (Aubl.) Willd., Sp. Pl. 2 (2): 1241. 1799. – *Tigarea aspera* Aubl., Hist. Pl. Guiane 2: 918. 1775. Type: French Guiana, Aublet s.n. (BM).

Tigarea aspera Aubl., Gui. 2: 918. 1775. Type: French Guiana: Aublet s.n (BM)

Liana, young branches and branchlets fasciculate pubescent, sparsely pilose to glabrescent when mature, older branches turning grayish with numerous cracks and fissures, bark brown, flaking off. Leaves coriaceous, depressed obovate or obovate, 3.4-6.8 × 3.6-4.6 cm, apex broadly acute or rotundate, emarginated, base petiolate, rotundate to truncate, margin simple, entire, sub-revolute; leaf blade sparsely scabrous to smooth adaxially, scabrous abaxially, with 5-8 pairs of lateral veins, venation brochidodromous; midrib and lateral veins scabrous, impressed adaxially, prominent abaxially; petioles 0.6-1 cm long, winged, ca. 0.5 mm wide and scabrous. Inflorescence terminal or axillary, thyrse multi-flowered, 4-6 cm long, the rachis sparsely appressed pubescent and stellate hairs. Flowers androdioecious. Bisexual flowers: sepals equal or nearly so, imbricate, ovate to obovate; outer sepals 2, 3 mm long; inner sepals 3, 6-7 mm long, scabrous externally, glabrous internally, the margins ciliated; petals 4-5, broadly ovate, white, 4-6 mm long. Staminate flowers: sepals unequal, scabrous externally, glabrous internally, the margin ciliated, outer sepals 2, lanceolate, 3-4 mm long, inner sepals 2, ovate, 5-6 mm long. Fruit 2-3 follicles, 4-9 mm long, glabrous, the style persistent, 1 mm long; seed dark, 1 per follicle, ca. 4 mm long; aril laciniate and covering one-third to one-half of the seed.

Distribution: Evergreen lowland forest up to 200 m alt. 15 collections studied, all from the Guianas: (GU: 3; SU: 1; FG: 11).

Selected specimens: Guyana: Mazaruni Station, Fanshawe F3405 (Record #6969 (NY, K). Suriname: Sipaliwini, vic. of Blanche Marie Waterfall on Nickerie R., Evan 2357 (US). French Guiana: vicinity of Cayenne, hill above Grant's Road, Matabon, Broadway 540 (US, NY); Cayenne, Sagot 1219 (P).

Vernacular names: Guyana: katuwai (AK)

6. **Tetracera volubilis** L. Sp. Pl. 1: 533. 1753. Neotype: Mexico, Veracruz, Zacuapan, Sulphur Spring, Purpus 2206 (F; isoneotype: US-840326; designated by Todzia & Barrie, Taxon 40: 652. 1991).

In the Guianas only: subsp. **volubilis**

Liana, young branches appressed pubescent to glabrescent when mature, older branches turning grayish with numerous cracks and fissures. Leaves chartaceous or coriaceous, obovate to oblanceolate, 9-16(-18) × 5-8 cm, apex rotundate, emarginated, base truncate, margin simple and entire; leaf blade glabrous on both sides, with 8-9 pairs of lateral veins, venation craspedodromous; midrib and lateral veins appressed pubescent, appressed adaxially, prominent abaxially; petioles 1.5-2 mm long, winged, ca. 1-2 mm wide and appressed pubescent. Inflorescence termial or axillary, thyrse many-flowered, 8-20 cm long, the rachis appressed pubescent mixed with stellate trichomes. Flowers androdioecious. Bisexual flowers: sepals unequal, imbricate, ovate to obovate, scabrous externally, sericeous internally, the margins ciliated, outer sepals 2-4 mm long, inner sepals 7-8 mm long; petals 4, broadly ovate, white to yellow, glabrescent on both sides, 3-4 mm long. Staminate flowers: sepals unequal, broadly ovate, papillate and sparely appressed pubescent externally, sericeous internally, the margins not ciliated, outer sepals 3, 1.5-3 mm long, inner sepals 2, 3-4 mm long; petals 5, obovate, white to cream, glabrous, 4-5 mm long. Fruit 3-4 follicles, 9-10 mm long, glabrous, the style persistent; seed dark, 1 per follicle, ca. 3 mm long, aril bright yellow to orange, laciniate and covering half the seed.

Distribution: Evergreen lowland forest up to 400 m alt. 3 collections studied, all from the Guianas: (GU: 2; FG: 1).

Selected specimens: Guyana: left bank upper Canje, ca. 35 Mi S of Torani Canal; 3 Mi SE of Digitima Cr., Pipoly, Gharbarran & Bacchus 9406 (US). French Guiana: Crique Plomb, Bassin du Sinnamary, Bordenave 833 (US).

Vernacular names: Guyana: katuwai (AK)

7. **Tetracera willdenowiana** Steud., Nomencl. Bot. ed. 2: 670. 1841 [1840]. – *Tetracera fagifolia* Willd. ex Schltdl., Linnaea 8: 174. 1833. Type: Brazil, Para, Hoffmannsegg s.n. (B, W, photo: F, G, HAL)

Tetracera surinamensis var. *reticulata* Lanj., Recueil Trav. Bot. Néerl., 37: 291. 1940. Type: Suriname: Upper Suriname R., Stahel 86 (U)

Scandent shrub or liana, young branches glabrous or nearly so, older branches with numerous cracks and fissures in the bark. Leaves coriaceous, obovate, ovate or elliptic, 4-12 × 4.4-8 cm, apex broadly acute to sub rotundate, base petiolate, rotundate or acute, margin simple, entire, revolute; leaf blade glabrous adaxially, scabrous abaxially, with (7-)11-15 pairs of lateral veins, venation brochidodromous; midrib appressed setose, glabrescent when mature; midrib and lateral veins slightly impressed adaxially, prominent abaxially; petioles (1.6-) 2.2-4.8 cm long; winged, ca. 0.5-1 mm sparsely pilose, glabrescent when mature. Inflorescence terminal and axilliary, thyrse many-flowered, 4-19 cm long, the rachis with a few stellate trichomes. Flowers androdioecious. Bisexual flowers 5-merous, sepals sub- to unequal, imbricate, widely obovate (orbicular), scabrous externally, glabrous internally, the margins ciliated, outer sepals 3, 4-5 mm long, inner ones 2, 4 - 5 mm long; petals 3-4, obovate, white, glabrous, ca. 5 mm long. Fruit 2-3 follicles, 6-7 mm long, glabrous, the style persistent, ca. 4 mm long; aril orange to yellow, laciniate and completely covering the seed.

KEY TO THE SUBSPECIES

1. Leaves obovate to widely ovate, apex truncate . . . .
. . . . .7a. *T. willdenowiana* subsp. *emarginata*
Leaves ovate to elliptic, apex slightly acute to acuminate . . . .
. . . . .7b. *T. willdenowiana* subsp. *willdenowiana*

7a. **Tetracera willdenowiana** subsp. **emarginata** Kubitzki Mitt. Bot. Staatssamml. München 8: 65. 1970. Type: Brazil: Pará. Belem, Mosqueiro, Sastre 153 (holotype: M, isotype: P).

Leaves coriaceous, obovate to widely ovate, 4-8.5 × 4.4-5.4 cm, apex truncate, emarginate, base petiolate, rotundate, rarely retuse; leaf blade with 7-10 pairs of lateral veins; petioles 1.4-1.6 cm long. Fruit 3-follicle, 3-4 mm long, style persistent ca. 3 mm long.

7b. **Tetracera willdenowiana** subsp. **willdenowiana**

Leaves ovate to elliptic, 4-12 × 4.4-8 cm, apex acuminate to acute, base petiolate to slightly acute; leaf blade glabrous or nearly so adaxially, scabrous abaxially, with 11-15 pairs of lateral veins; petiole 1.6-4.8 cm long.

Distribution: Evergreen lowland forest 50-200 m. 2 collections studied, all from the Guianas: (GU: 1; FG: 1).

Selected specimens: Guyana: Essequibo Island – Demerara: Arawak Amerindian land, Timerhead Resort, 3 km up Pokerero River from Santa Mission; NE of Ants Creek, Hoffman 880 (US). French Guiana: Haut Oyapock, Mt. St. Marcel, le long de al crique Camp Poivre, Sastre 4537 (CAY)

Vernacular names: Guyana: katuwai (AK)

117. VITACEAE

by

JULIO A. LOMBARDI [4]

Lianas, rarely shrubs or treelets; plants monoecious or rarely dioecious; shoots of sympodial development, vegetative branches of indefinite development, reproductive branches short, unbranched, with or without leaves; tendrils usually present, leaf-opposed, rarely axillary. Leaves stipulate, stipules small, caducous or less often persistent; petiolate; alternate, palmately to pinnately lobed or compound, less often simple. Inflorescences leaf-opposed, axillary or extra-axillary, pedunculate, composite, cymose or racemose. Flowers pedicellate or sessile, actinomorphic, bisexual or functionally unisexual (in polygamous species); sepals 4-5, mostly connate; petals 4-5, free or connate at base or apex, valvate, caducous at anthesis, rarely persistent in fruit; androecium apostemonous, stamens 4-5, epipetalous, anthers (2-)4-locular; disc when present intrastaminal, free or adnate to ovary wall, composed of 5 free glands or annular and 4-5(-10) sulcate or lobed; ovary superior, complete or incomplete 2-locular, placentation axillary, ovules anatropous, ascendant, 2 per locule, style central, 1 or rarely absent, stigma apical, 1, punctual, discoid, capitate or rarely 4-lobed. Fruits berries or rarely amphisarcum, with 1-2(-4) seeds; seeds with thin sarcotesta, tipically with two ventral intrusions into endosperm (foveae) laterally to a prominent raphe, sometimes with an abaxial chalazae, endosperm copious, commonly 3-lobed or ruminate, embryo minute.

Distribution: Worldwide, consisting of 13-14 genera and about 600 species, the great majority in tropical and subtropical climates; 4 genera with 82 species known from the Neotropics, 1 genus and 14 species in the Guianas.

LITERATURE

Baker, J.G. 1871. Ampelideae. In C.F.P. von Martius, Fl. Bras. 14(2): 197-220.
Cambessèdes, J. 1828. Ampelideae. In A.F.C.P. Saint-Hilaire, Fl. Bras. Merid. 1: 342-347.

4 Departamento de Botânica, Instituto de Ciências Biológicas, Universidade Federal de Minas Gerais, Av. Antônio Carlos 6627, 31270-110, Belo Horizonte, Minas Gerais, Brazil.
The author has received grants from the Universidade Federal de Minas Gerais through the Pró-Reitoria de Pesquisa (numbers 23072.050311/94-99, 23072.040763/95-25 and 23072.044225/96-36), and CNPq (research fellowship grant).

42

Descoings, B. 1991. Contribution à l'étude des Vitacées d'Amérique tropicale: deux Cissus_nouveaux des Guyanes. Bull. Soc. Bot. France, Lett. Bot. 138: 249-256.

Descoings, B. 1994. Contribution à l'étude des Vitacées d'Amérique tropicale. II. – Cissus haematantha Miq. Acta Bot. Gallica 141: 361-366.

Descoings, B. 2008. Contribution à l'étude des Vitacées d'Amérique tropicale. III - Cissus spinosa, C. subrhomboidea, C. flavens et C. kawensis, spp. nov. J. Bot. Soc. Bot. France 44: 3-17.

Descoigns, B. 2011. Contribution à l'étude des Vitacées d'Amérique tropicale. IV - Cissus erosa, C. sicyoides et C. ulmifolia. J. Bot. Soc. Bot. France 53: 37-54.

Gilg, E. 1896. Vitaceae. In A. Engler & K. Prantl, Nat. Pflanzenfam. 3(5): 427-456.

Görts-van Rijn, A.R.A. 1979. Vitaceae. In A.L. Stoffers & J.C. Lindeman, Flora of Suriname 5(1): 335-343.

Heald, S.V. 2002. Vitaceae. In S.A. Mori et al., Guide to the Vascular Plants of Central French Guiana. Mem. New York Bot. Gard. 76(2): 740-742.

Lombardi, J.A. 1995. Typification of names of South American Cissus (Vitaceae). Taxon 44: 193-206.

Lombardi, J.A. 2000. Vitaceae. Gêneros Ampelocissus, Ampelopsis e Cissus. Fl. Neotrop. Monogr. 80: 1-251.

Nicolson, D.H. & C. Jarvis. 1984. Cissus verticillata, a new combination for C. sicyoides (Vitaceae). Taxon 33: 726-727.

Planchon, J.E. 1887. Monographie des Ampélidées vraies. In A.L.P.P. & C. De Candolle, Monogr. Phan. 5(2): 305-654.

Roosmalen, M.G.M. van. 1985. Vitaceae. In Fruits of the Guianan Flora, p. 434-436. Utrecht University, Institute of Systematic Botany, Utrecht.

Suessenguth, K. 1953. Vitaceae. In A. Engler & K. Prantl, Nat. Pflanzenfam. ed. 2. 20d: 174-333.

CISSUS L., Sp. Pl.: 117. 1753.
Type: *C. vitiginea* L.

Lianas, rarely shrubs; plants monoecious; tendrils leaf-opposed. Leaves simple or compound; leaf shape variable in a single specimen, sometimes vegetative and reproductive branches showing different leaf shape, or leaves absent in reproductive branches; tertiary veins usually conspicuous, pearl glands present in young parts, rarely persistent and conspicuous in mature parts. Inflorescences leaf-opposed, compound cymes, flat-topped, rarely umbel-like or glomerule-like and with convex apex or elongate;

bracts scale-like, minute. Flowers pedicellate, bisexual, possibly protrandric; calyx cup-shaped, calyx lobes 4(-5); petals 4(-5), margins coherent, caducous at anthesis, rarely persistent in the fruit; stamens 4(-5), base of filaments adherent to disc, anthers 2-locular, dehiscence longitudinal; disc annular, covering ovary wall and adnate to it, ovary apex rarely free; stigma minute or slightly capitate. Fruit a berry or rarely an amphisarcum with epicarp thin and papery; seeds 1(-4), pear-shaped or rhombic, symmetric or asymmetric, laterally rounded or flattened.

Distribution: 63 species in tropical S America, 14 species known from the Guianas.

KEY TO THE SPECIES

1 Plants with simple leaves only 2
  Plants with trifoliolate leaves, at least in vegetative shoots 5

2 Leaves densely papillose abaxially, drying ochre or yellow ....
  .... 3. *C. descoingsii*
  Leaves not papillose abaxially, nor drying ochre or yellow 3

3 Leaves membranous, drying blackish................... 10. *C. tinctoria*
  Leaves chartaceous or carnose, not drying blackish................. 4

4 Flower buds conical; leaves with tertiary veins inconspicuous ....
  .... 13. *C. venezuelensis*
  Flower buds ellipsoidal; leaves with tertiary veins conspicuous ....
  .... 14. *C. verticillata*

5 Plants armed .... 8. *C. spinosa*
  Plants unarmed 6

6 Branched (malpighiaceous) hairs present, at least at the peduncle apex,
  glandular hairs absent... 7
  Branched hairs absent, at least at the peduncle apex, glandular hairs present
  or absent 8

7 Leaves absent in reproductive shoots when opposite to inflorescences .....
  .... 2. *C. amapaensis*
  Leaves present in reproductive shoots when opposite to inflorescences, at
  least in the proximal part of shoot .... 5. *C. erosa*

8 Reproductive branches in the apex without leaves, or these very reduced, or
  with very small or absent lateral leaflets 9
  Reproductive branches with smaller leaves, but these are not notably
  reduced 11

9 Plants drying tile-red, reproductive branches cylindrical ....
  .... 6. *C. haematantha*

Plants when dried not tile-red, reproductive branches sulcate, angled or winged 10

10 Inflorescences 2.3-5.3 × 1.7-4 cm; peduncles 0.6-2.1 cm long, fruits ca. 7-9 mm × 5-7 mm..1. *C. alata*
Inflorescences 6.1-9.4 × 5-8.3 cm; peduncles 2.8-5.2 cm long, fruits ca. 27 x18 mm...7. *C. nobilis*

11 Petals glabrous ....12. *C. ulmifolia*
Petals pubescent 13

12 Leaflet margin denticulate and sinuate, crenate or lacerate, sometimes undulate; mature berry yellowish-green...............4. *C. duarteana*
Leaflet margin denticulate; mature berry purple....... 9. *C. surinamensis*

1. **Cissus alata** Jacq., Selec. Stirp. Amer. Hist.: 23. 1763. – *Vitis alata* (Jacq.) Kuntze, Revis. Gen. Pl. 3: 40. 1898. Lectotype (designated by Lombardi 1995): Guyana, as "Carthagenae", unknown collector s.n. (hololectotype BM (only leaf on top right of sheet)).

Cissus rhombifolia Vahl, Eclog. Amer. 2: 10. 1798. – *Vitis rhombifolia* (Vahl) Baker in Mart., Fl. Bras. 14(2): 207. 1871. Neotype (designated by Lombardi 1995): Trinidad, Nariva swamp, Kalloo 1196 (holoneotype TRIN, isoneotype L).

Liana; trichomes unbranched, glandular and eglandular. Branches sulcate, angled or rarely winged, reproductive branches without leaves at apex or with leaves with reduced lateral leaflets. Stipules spathulate or falcate, 4-7.5 × 1-2.5 mm, gibbous, pilose, reflexed. Leaves trifoliolate; petiole 1.1-12 cm long; petiolule 0-1 cm long; leaflet blade chartaceous, elliptic, subelliptic, rhombic or subovate, the central one (2.3-)12-17.7 × (0.8-)8.3-11.1 cm, the lateral ones (0.4-)8.1-14.8 × (0.3-)4.7-10.5 cm, margin denticulate, rarely toothed or crenulate, apex acute or obtuse, base oblique or attenuate, scabrous adaxially, puberulent, subcanescent or sparse tomentose along veins abaxially, glabrescent. Inflorescence 2.3-5.3 × 1.7-4 cm; peduncle 0.6-2.1 cm long, tomentose or puberulent, greenish; pedicels 2-4 mm long, puberulent or pubescent, greenish. Flower buds conoidal; perianth reddish or greenish; calyx 1-1.5 × 2 mm, truncate, base rounded or truncate, puberulent; corolla in bud 1.5-2.5 × 1.5-2 mm, petals glabrous or sometimes sparse puberulent, caducous; anther dehiscence latrorse; disc orange. Berry purple, globose or pear-shaped, ca. 7-9 mm × 5-7 mm, smooth; seed 1, subturbinate, flattened, 6-7 × 5-6 mm, laterally rugose.

Distribution: Mexico, Panama, Colombia, Venezuela, Trinidad, Guyana, Ecuador, Peru and Bolivia; over 90 collections studied, 1 from the Guianas (GU: 1).

Specimen examined: Guyana: Kanuku Mts., Moco Moco R., Jansen-Jacobs *et al.* 4490 (BHCB, U, P).

Phenology: Flowering reported from April and July; fruiting not recorded for the Guianas.

Note: The type specimen of *Cissus alata* is mixed with fragments of *Cissus trigona* Willd. ex Schult. & Schult. f.

2. **Cissus amapaensis** Lombardi, Novon 6: 195. 1996. Type: Brazil, Amapá, Mazagão, Pires & Silva 1234 (holotype NY).

Cissus flavens Desc., J. Bot. Soc. Bot. France 44: 11. 2008. Type: French Guiana, fleuve Oyapock, savane roche "Canari Zorzo", Oldeman B-2501 (holotype CAY (image seen)).

Liana; trichomes malpighiaceous, eglandular. Branches cylindric, apex of reproductive branches without inflorescence-opposite leaves. Stipules triangular, 2-3 × 1.5-3.5 mm, glabrous, ciliate. Leaves trifoliolate; petiole 3.9-4.5 cm long; petiolule 0.3-1.4 cm long; leaflet blade chartaceous, elliptic or subelliptic, the central one 11.5-12 × 3.3-3.4 cm, the lateral ones 7.9-9.4 × 2.2-2.7 cm, margin denticulate, apex acute or caudate, base cuneate, glabrous adaxially, sparse pubescent along veins abaxially. Inflorescence 3.2 × 2.1-2.6 cm; peduncle 1.5-1.7 cm long, sparse pubescent at apex, probably red; pedicels 3 mm long, glabrous, red. Flower buds ellipsoidal; perianth red; calyx 1.5 × 2 mm, truncate, base rounded, glabrous; corolla in bud 1.5 × 1.5 mm, petals glabrous, papillose, caducous; anther dehiscence extrorse. Fruit and seeds not seen.

Distribution: Brazil (Brazilian Amazon), French Guiana; 2 collections studied, 1 from the Guianas (FG: 1).

Phenology: Flowering reported in June (Brazil, Amapá State).

Note: Known only by the 2 cited type specimens.

3. **Cissus descoingsii** Lombardi, Candollea 51: 370. 1996. – *Cissus guyanensis* Desc., Bull. Soc. Bot. France, Lett. Bot. 138: 249-256. 1991, nom. illeg. Type: French Guiana, Montagne de Kaw, de Granville 7277 (holotype U, isotypes B, CAY (not seen), NY (not seen), P). – Fig. 1

Cissus kawensis Desc., J. Bot. Soc. Bot. France 44: 14. 2008. Type: French Guiana, Montagne de Kaw, Piste Roura/Kaw, PK 40, Feuillet 2925 (holotype CAY (image seen), isotypes U, P).

46

Liana; trichomes unbranched and malpighiaceous, eglandular. Branches cylindric or 4-angled. Stipules triangular, subovate or rhombic, 2-4 × 1-4 mm, glabrous or sparse sericeous. Leaves simple; petioles (0.7-)1.1-19.4 cm long; leaf blade chartaceous, cordiform, elliptic, ovate, triangular or oblong, 5.1-17.3 × 2.3-16.5 cm, margin denticulate, apex acute or acuminate, base cordate, subcordate, rounded, cuneate or truncate, densely papillose and drying ochre abaxially, sparse sericeous on both sides, glabrescent. Inflorescence 2.2-6.2 × 2.5-6.9 cm; peduncle 0.3-3.2 cm long, puberulent, red or green; pedicels 1-4 mm long, sericeous, glabrescent, greenish or red. Flower buds spindle-shaped; perianth yellowish or red; calyx 0.5-1 × 1-2 mm, truncate, base rounded, sericeous,

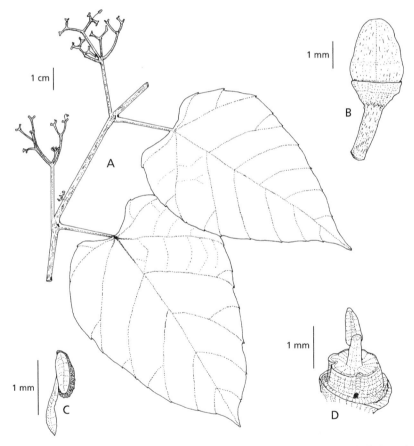

Fig. 1. *Cissus descoingsii* Lombardi: A, reproductive branch; B, flower bud; C, lateral view of stamen; D, disc view from above, petals and 3 stamens removed; (A-D, Andel *et al.* 1133).

glabrescent; corolla in bud 1-3 × 1-2 mm, petals glabrous or sericeous, caducous; anther dehiscence latrorse; disc yellow. Berry purple, ellipsoid, 13-20 xx 6-9 mm, with sparse lenticels; seed 1, spindle-shaped, rounded, 12.5 × 5-7 mm, laterally grooved.

Distribution: Costa Rica, Panama, Colombia, Venezuela, Guyana, French Guiana, Ecuador, Peru and Brazil; over 90 collections studied, 17 from the Guianas (GU: 5; FG: 12).

Selected specimens: Guyana: Blue Mt., Kokerite, Barama R., van Andel *et al.* 1133; Kariako, Barama R., van Andel *et al.* 1657 (BHCB, U). French Guiana: Mt. de Kaw, Cremers *et al.* 9760; Mt. Favard, Mt. de Kaw, Hoff *et al.* 6326 (BHCB, CAY, P).

Phenology: Flowering reported from May, September and December, fruiting from February.

Notes: *Cissus guyanensis*, was invalidly published by Descoings (1991), without specific indication of a holotype.

Some plants outside the Guianas differ from the typical *C. descoingsii* by a dense hair-cover, however the specimens are insufficient distinct to be segregated in another taxon.

4. **Cissus duarteana** Cambess. in A. St.-Hil., Fl. Bras. Merid. 1: 343. 1828. – *Vitis duarteana* (Cambess.) Baker in Mart., Fl. Bras. 14(2): 204. 1871. Type: Brazil, Minas Gerais, "campis prope vicum Contendas", Saint-Hilaire 139 (holotype MPU (not seen), isotype P = F Neg 35986). – Fig. 2

Shrub or liana; trichomes unbranched and eglandular. Branches angled or sulcate. Stipules triangular or falcate, 3-7 × 2-4 mm, sparse villous, ciliate. Leaves trifoliolate, occasionally some of the leaves simple and irregularly 3-lobed; petioles (0.4-)2.3-7.8 cm long; petiolule 0-0.5 cm long; leaflet blade chartaceous, rhombic, obtrullate, elliptic, subrhombic or subobovate, the central one (5-)12.8-19.8 × (1.6-)4.5-10.1 cm, the lateral ones (1.7-)3.6-11.8 × (1-)1.4-5.8 cm, margin ciliate, denticulate and sinuate, crenate or lacerate, sometimes undulate, rarely lobed, apex acute, base attenuate, strigose, subvillous or tomentose, glabrescent, rarely glabrous. Inflorescence 4.2-7.4 × 2.1-3.3 cm; peduncle 2.1-5.3 cm long, villous, reddish; pedicels 1-3 mm long, velutinous or subvelutinous, reddish or green, bend in fruit. Flower buds conoidal; perianth cream-yellowish or pinkish; calyx 1 × 1-1.5 mm, truncate, base rounded, puberulent or sparse pilose; corolla in bud 1-1.5 × 1-1.5 mm, petals puberulent or sparse pilose, caducous; anther dehiscence extrorse. Berry

yellowish-green, globose or subglobose, ca. 8 × 5 mm, smooth; seeds
1(2), subturbinate, rounded, ca. 6 × 4 mm, laterally sulcate or smooth.

Distribution: Guyana, Suriname, French Guiana, Brazil, Bolivia and
Paraguay; over 90 collections studied, 16 from the Guianas (GU: 2; SU:
5; FG: 9).

Selected specimens: Guyana: Rupununi Distr., Shea Village, Jansen-
Jacobs *et al.* 4796 (BHCB, CAY, K, U, MO, P); Karanambo, Maas *et
al.* 7258 (B, CAY, U, MO, US). Suriname: Sipaliwini savanna area on
Brazilian frontier, Oldenburger *et al.* 809 (U); Sipaliwini R., Rombouts
441 (A, U). French Guiana: Savanne Morpio, entre Morpio et Roches

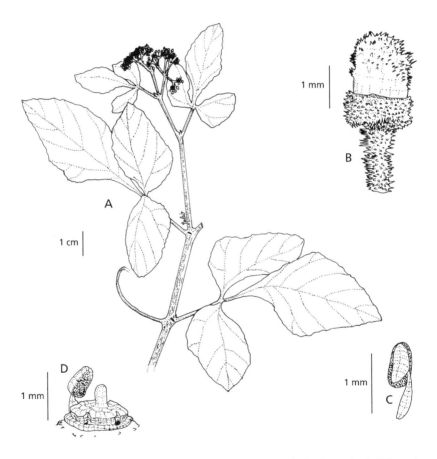

Fig. 2. *Cissus duarteana* Cambess.: A, reproductive branch; B, flower bud; C, lateral
view of stamen; D, disc view from above, petals and 3 stamens removed; (A, Cremers
14581; B-D, Prévost 2131).

Blanches, Cremers 14581 (BHCB, CAY, P, U, US); Savanne Mamaribo, Hoff & Cremers 5607 (CAY, U, US); Savanne au bord de la Piste de St. Elie, Prévost 2131 (BHCB, CAY, K).

Phenology: Flowering reported from July to September, and January to April, fruiting reported in August, September, January, and from March to May.

Notes: In the savannas, *Cissus duarteana*, similarly to *Cissus erosa*, frequently grows at the start of the growing season as a shrub with great lobed or trifoliolate leaves and short petioles, more rarely as a liana with smaller trifoliate leaves and longer petioles. Due to this similarity on the growth form these species can be confused, but C. erosa has red flowers and branched hairs, while C. duarteana has cream-yellowish or pinkish flowers and unbranched hairs.this species was identified by Descoings (2008) as *C. subrhomboidea* (Baker) Planch., but the two species can be distinguished by the leaflets margins (denticulate and sinuate, crenate or lacerate in *C. duarteana* vs. only denticulate in *C. subrhomboidea*), the mature fruit (green vs. purple), and the seed morphology (rounded vs. flattened). Additionally, *C. subrhomboidea* occurs only in southeastern and Brazil and adjacent Paraguay.

5. **Cissus erosa** Rich., Actes Soc. Hist. Nat. Paris 1: 106. 1792. – *Vitis erosa* (Rich.) Baker in Mart., Fl. Bras. 14(2): 210. 1871. Type: French Guiana, Leblond 78 (holotype G (not seen) = F Neg 23806, isotype F).

Cissus lucida Poir. in Lam., Encycl. Suppl. 1: 106. 1810. Type: French Guiana, Cayenne, Martin s.n. (holotype P-LAM (not seen)).
Cissus salutaris Kunth in Humb., Bonpl. & Kunth, Nov. Gen. Sp. 5: 225. 1822. – *Vitis salutaris* (Kunth) Baker in Mart., Fl. Bras. 14(2): 211. 1871. – *Cissus erosa* Rich. var. *salutaris*_(Kunth) Planch. in A. DC. & C. DC., Monogr. Phan. 5: 548. 1887. Type: Venezuela, "prope Quetepe et Cumana", Humboldt & Bonpland s.n. (holotype P (not seen) = F Neg 35991).
Cissus elongata Miq., Stirp. Surinam. Select. 110. 1851 (non Roxb., Fl. Ind. 1: 429. 1820), nom. illeg. – *Vitis miqueliana* Baker in Mart., Fl. Bras. 14(2): 210. 1871. Type: Suriname, Hostmann 1301 (holotype K (not seen) = cibachrome at UEC, isotype BM, G (not seen) = F Neg 23807, K (not seen) = cibachrome at UEC, W).
Cissus erosa Rich. subsp. *linearifolia* (Baker) Lombardi, Taxon 44: 196. 1995. – *Vitis miqueliana*_Baker var. *linearifolia*_Baker in Mart., Fl. Bras. 14(2): 211. 1871. – *Cissus erosa*_Rich._var. *linearifolia* (Baker) Hoehne, Indice Bibl. Num. 273. 1951. Lectoype (designated by Lombardi 1995): Brazil, Tocantins, Natividade, Gardner 3076 (hololectotype K), syn.nov.

Liana; trichomes unbranched and malpighiaceous, eglandular. Branches 4-6-angled, winged or cylindric, reproductive branches might show smaller leaves, but not notably reduced. Stipules triangular, rhombic or oblong, 2-3 × 1-2 mm, puberulent or sparse tomentose. Leaves trifoliolate; petiole (0.1-)3.5-14 cm long; petiolule (0-)0.5-2.5 cm long; leaflet blade chartaceous, obovate, elliptic, rhombic, oblong, subobovate or lanceolate, the central one (3.7-)4.3-19.9 × (1.5-)4.8-13.9 cm, the lateral ones 3.1-14.7 × (0.4-)4.5-11.7 cm, margin denticulate or toothed, rarely lobed, apex acute, base attenuate, cuneate, oblique, truncate or rounded, glabrous, puberulent or hispid. Inflorescence 5.8-20.5 × 2.3-10.1 cm; peduncle 3-17.8 cm long, sericeous, glabrescent, green or reddish; pedicels 1.5-3 mm long, sericeous, glabrescent, red. Flower buds ellipsoidal; perianth red outside, orange inside; calyx 0.5-1 × 1-2 mm, with rounded lobes or truncate, base rounded, sericeous; corolla in bud 1-2 × 1-2 mm, petals glabrous, caducous; anther dehiscence latrorse. Berry purple, globose or pear-shaped, 7-7.5 × 5-7 mm, smooth; seed 1, subturbinate, rounded, ca. 5-7 × 4-5 mm, laterally sulcate or smooth.

Distribution: Mexico to southeastern Brazil and Paraguay, including Caribbean; over 1000 collections studied, 199 from the Guianas (GU: 50; SU: 80; FG: 69).

Selected specimens: Guyana: Wanama R., Northwest Distr., de la Cruz 3974 (F?); Pomeroon R., Pomeroon Distr., de la Cruz 3098 (F); Mahaica R., St. Cuthbert's Mission, Davis 30 (NY); Mazaruni R., Jenman 5352 (BM, F); NW slopes of Kanuku Mts., Moku-Moku Cr., Smith 3520 (B, F, U, P, S, W). Suriname: Upper Suriname R., BW 5272 (U, UC); along Paulus Cr., lower Suriname R., Mennega 212 (C, U); forest of Zandery, Samuels 255 (A, U, P); pr. urbem Paramaribo, Kappler 1598 (U, P, S, W, Z); Tafelberg, vicinity Saron Cr., Saramacca R., Maguire 23769 (F, MG, S). French Guiana: bassin du Bas-Oyapock, savanne Roche du 14 Juillet, Cremers 12193 (B, CAY, U, P); piste du Village Eskol, Mt. de Kaw, Le Goff & Hoff 225 (BHCB, CAY, P); Roche Koutou, bassin du Haut-Marouini, de Granville et al. 9304 (B, U, P); Cr. Cabassou, Ile de Cayenne, Hoff 5405 (BHCB, CAY); Maripa, Sauvain 338 (B, U, P).

Vernacular names: Surinam: boen-ati-mama (Sranan, van Roosmalen, 1985), lebi-kifaia (Paramaccan, van Roosmalen, 1985), rode-kop and wilde-napie (Surinamese Dutch). French Guiana: agui-fassar (Créole), amatalea (Wayãpi), amataleasili (Wayãpi), kifaya (Boni, Taki-taki), kii-faja (Boni), lebi kii faya (Aluku), lébi-ki-faia (Taki-Taki), liane-ravet (Créole), nap-kayma (Wayana), shakacarib (Palikur), sui-tõgo (Saramaccan).

Phenology: Flowering and fruiting reported along the year.

Notes: *Cissus erosa* is a very variable species that grows in the central

Brazilian savannas ('cerrados') as perennial shrub with aerial branches that dye off every dry season. In this habitat, the plants show innitially big simple, trilobed or trifoliolate leaves with short petioles, looking similar to *C. duarteana*, and later the leaves become smaller and trifoliolate.

Specimens with lanceolate leaflets were determined by Lombardi (2000) as a distinct subspecies, *C. erosa* Rich. subsp. *linearifolia* (Baker) Lombardi, but there are many intermediate forms.This subspecies is here placed as synonym of *C. erosa* Rich.

Pilose forms had received specific status (e.g. *C. salutaris* Kunth) but occur sporadically in the whole geographic range of the species.

6. **Cissus haematantha** Miq., Linnaea 26: 220. 1854. Type: Suriname, Maipuri Kreek, Kappler 1959 (holotype U (not seen), isotypes C, GOET (not seen) = photo at BHCB, S). – Fig. 3

Liana, drying tile-red; trichomes unbranched and eglandular, mixed with glandular ones. Reproductive branches cylindric, with leaves with very reduced or absent lateral leaflets, or leaflets united, vegetative branches 4-angled or winged. Stipules deltate or subfalcate, 4-6 × 2-9 mm, slightly gibbous, puberulent, reflexed. Leaves trifoliolate; petiole (0.4-)4.2-14.1 cm long; petiolule 0-0.3(-1.7) cm long; leaflet blade chartaceous, elliptic, lanceolate, subovate or ovate, the central one (2.9-)7.1-11.8 × (0.7-)3.1-6.7 cm, the lateral ones (0.7-)3.6-10.3 × (0.4-)1.8-5.2 cm, margin denticulate, apex acute or acuminate, base attenuate or rounded, glabrous or slightly scabrous. Inflorescence 2.4-6.3 × 2.4-5.7 cm; peduncle 1-2.7 cm long, puberulent or scabrous, probably red; pedicels 2.5-3 mm long, puberulent, red. Flower buds conoidal; perianth red or orange; calyx 1-1.5 × 2 mm, truncate or irregularly lobed, base truncate and irregularly lobed, glabrous or puberulent at base; corolla in bud 1-2 × 2 mm, petals glabrous, papillose, caducous; anther dehiscence extrorse. Berry purple, ellipsoid, ca. 17 × 12 mm, with sparse lenticels; seed 1, prisma-shaped, flattened, ca. 13 × 8 mm, laterally rugose.

Distribution: Venezuela, Trinidad, Guyana, Suriname, French Guiana and Brazil (Roraima); 30 collections studied, 23 from the Guianas (GU: 3; SU: 4; FG: 16).

Selected specimens: Guyana: Upper Essequibo R., Rewa R., near Corona Falls, Jansen-Jacobs *et al.* 5799 (BHCB, U, P); Upper Takutu-Upper Essequibo, Kassakaityu R., 25 km upstream from juncture with Essequibo R., Henkel 5316 (MO). Suriname: Nickerie, area of Kabalebo Dam project, Matapi, behind levee of Corantyne R., Heyde & Lindeman

108 (U). French Guiana: piste de Petit Saut, bassin du Sinnamary, Cremers *et al*. 12440 (BHCB, CAY); piste de la Mt. Cacao, bassin de la Comté, Hoff *et al*. 6277 (BHCB, CAY); R. Camopi, Yaniou, Oldeman B-1429 (CAY, U, P, US); forêt Paracou, Site Expérimental CTFT, Riéra 1465 (BHCB, CAY).

Vernacular name: French Guiana: yãwĩlɛmɔ (Wayãpi).

Phenology: Flowering reported from January to March, and October, fruiting in April and September.

Note: *Cissus haematantha* is easily distinguished by its strong red color when dried.

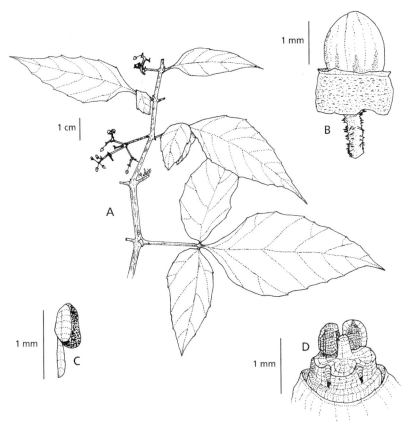

Fig. 3. *Cissus haematantha* Miq.: A, reproductive branch; B, flower bud; C, lateral view of stamen; D, disc view from above, petals and 2 stamens removed; (A-D, Riéba 1465).

7. **Cissus nobilis** Kuhlm., Anais Reunião Sul-Amer. Bot. 1(3): 83.
1938. Lectotype (designated by Lombardi 1995): illustration in Anais
Reunião Sul-Amer. Bot. 1(3): 11. 1938.

Liana; trichomes unbranched and eglandular, or mixed with glandular
ones. Branches 4-winged, sometimes reproductive branches with lateral
leaflets very reduced or leaflets united. Stipules deltate, 2-6 × 2-5 mm,
glabrous or puberulent, ciliate. Leaves trifoliolate; petiole 1.4-16.7
cm long; petiolule 0-1.1 cm long; leaflet blade chartaceous, elliptic,
subovate or subelliptic, the central one (3.6-)11.5-22 × (1.3-)4-17.2 cm,
the lateral ones (0.5-)6-15.2 × (0.2-)3.3-13.1 cm, margin denticulate,
apex acute or acuminate, base cuneate, attenuate, rounded or oblique,
glabrous adaxially, hispid abaxially, mainly along veins. Inflorescence
6.1-9.4 × 5-8.3 cm; peduncle 2.8-5.2 cm long, sparse hispid at base and
puberulent or pulverulent at apex, green; pedicels 3-5 mm long, sparse
pilose, greenish. Flower buds conoidal; perianth greenish or purplish;
calyx 1-2 × 3-4 mm, truncate, base truncate, laterally enlarged and
discoid or irregularly lobed, sparse pubescent mainly at base; corolla in
bud 1-2 × 2 mm, petals papillose, caducous; anther dehiscence extrorse,
disc yellowish. Berry purple, ellipsoid, ca. 27 × 18 mm, smooth; seed 1,
prisma-shaped, flattened, ca. 20 × 11 mm, laterally smooth.

Distribution: Suriname, Peru and Brazil; 19 collections studied, 1
from the Guianas (SU: 1).

Specimen examined: Suriname: Brownsberg, Plotkin Z-05606
(ECON).

Vernacular name: Suriname: bon'ati-mama.

Phenology: Not known from the Guianas.

Note: *Cissus nobilis* occurs mainly in the forests of the Atlantic coast of
Brazil. The specimens from Peru, Suriname and Brazilian Amazon are
very fragmentary collections and more pubescent than those from the
Atlantic forest, perhaps the former ones represent an undescribed taxon.

8. **Cissus spinosa** Cambess. in A. St.-Hil., Fl. Bras. Merid. 1: 345. 1828.
– *Vitis spinosa* (Cambess.) Baker in Mart., Fl. Bras. 14: 204. 1871.
Type: Brazil, Minas Gerais, Mengahi, Laruotte s.n. (holotype P (not
seen) = F Neg 35998, isotype F).   – Fig. 4

Vitis parkeri Baker in Mart., Fl. Bras. 14(2): 209. 1871. – *Cissus parkeri*
(Baker) Planch. in A. DC. & C. DC., Monogr. Phan. 5: 550. 1887.
Lectotype (designated by Lombardi 1995): Guyana, Demerara, Parker s.n.
(hololectotype K). Syntype: Suriname, Salem, Wullschlägel 66 (BR).

Liana; trichomes malpighiaceous mixed with unbranched ones, eglandular. Branches cylindric, old ones armed with turbinate emergences. Stipules elliptic, 3-5.5 × 2-6 mm, gibbous, puberulent, ciliate, gibbous center apparently glandular, drying blackish. Leaves trifoliolate; petiole (1.1-)3.5-11.3 cm long; petiolule (0.1-)0.4-2.1 cm long; leaflet blade chartaceous, rhombic, subcircular or subobovate, the central one (4.1-)5.4-15.9 × (2.2-)6.8-9 cm, the lateral ones (4.4-)9-14.5 × (1.7-)6.2-8.4 cm, margin denticulate or toothed, sometimes lobed, apex acute, base attenuate or oblique, both sides velutinous, sparse sericeous, sparse pubescent and glabrescent adaxially, pubescent or pulverulent and glaucous or ferruginous abaxially. Inflorescence 9.2-17 × 4.5-12.7 cm; peduncle 6.7-12.7 cm, sericeous mainly at apex, red; pedicels 1.5-3.5 mm long, tomentose or subtomentose, bend in fruit, red. Flower buds conoidal; perianth red outside, orange inside; calyx 0.5-1 × 1-2 mm, truncate or with rounded lobes, base rounded, subtomentose at base, papillose; corolla in bud 1-2 × 1-2 mm, petals sometimes sparse pilose, papillose, caducous; anther dehiscence latrorse; disc yellow. Berry purple, ellipsoid, ca. 10 × 5.5 mm, smooth; seeds 1(-2), subturbinate, rounded, ca. 9 × 6 mm, laterally smooth.

Distribution: Venezuela, Guyana, Suriname, French Guiana, Brazil, Bolivia and Paraguay; over 200 collections studied, 28 from the Guianas (GU: 5; SU: 11; FG: 12).

Selected specimens: Guyana: Lesbeholden, Black Bush Polder, Persaud 231 (U); along road from Cane Grove to Lama Conservancy, Hahn 3812 (U). Suriname: Fernandesweg, Jonker-Verhoef & Jonker 46 (U); inter Coppename fluv. ostium et Coronie oppidum, Lanjouw & Lindeman 1485 (U). French Guiana: savanne de Cr. Jacques, 10 km W of Mana, Cowan 38883 (F, US); Ilet Maripa, Route de l'Anse, Le Goff & Hoff 127 (BHCB, CAY, P); Route de Kaw, Mt. de Kaw, de Granville 9160 (B, CAY, U, MO, P); Cr. Canceler, Toriola-Marbot & Hoff 213 (BHCB, CAY).

Vernacular names: Guyana: snake vine. Suriname: boesie watramon. French Guiana: pri-pri.

Phenology: Flowering and fruiting reported along the year.

Note: *Cissus spinosa* is always associated to rivers or water bodies and probably dispersed by fishes.

9. **Cissus surinamensis** Desc., Bull. Soc. Bot. France, Lett. Bot. 138: 252. 1991. Type: Suriname, area of Kabalebo Dam project, Distr. Nickerie, Lindeman & Görts-van Rijn *et al.* 457 (holotype U, isotypes BHCB, NY).

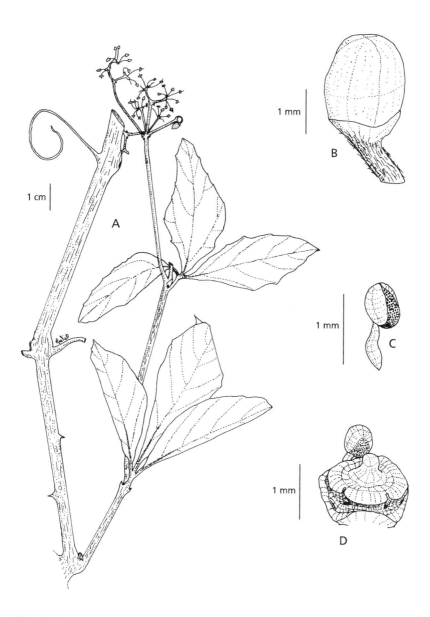

Fig. 4. *Cissus spinosa* Cambess.: A, reproductive branch; B, flower bud; C, lateral view of stamen; D, disc view from above, petals and 3 stamens removed; (A-D, Le Goff & Hoff 127).

Liana; sparse pilose or hispid, trichomes unbranched, eglandular and glandular. Branches cylindric, old ones ramentaceous with sparse short tubercles. Stipules falcate, 2-3 × 0.5-2 mm, gibbous, glabrous. Leaves trifoliolate; petiole 2.2-9.1 cm long; petiolule 0-1 cm long; leaflet blade chartaceous, elliptic, subelliptic or triangular, the central one (2.9-)4-14.9 × (0.7-)1.8-7.7 cm, the lateral ones (1-)2.4-9 × (0.8-)1.5-4.9 cm, abaxial surface sometimes reddish, margin denticulate, rare lobed, apex acute, base attenuate, cuneate or oblique, sparse pilose adaxially, puberulent or hispid abaxially. Inflorescence 2.2-2.8 × 1.7-2.8 cm; peduncle 1.2-1.6 cm long, hispid, green; pedicels 1.5-2 mm long, puberulent or hispid, greenish. Flower buds conoidal; perianth yellowish; calyx 1 mm. long, 1.5-2 mm, lobes deltate or rare acute, base commonly truncate or sometimes lobed, puberulent; corolla in bud 1 × 1-1.5 mm, petals puberulent or tomentose, glabrescent, caducous; anther dehiscence extrorse. Berry purple, globose or pear-shaped, ca. 10 × 8 mm, with sparse lenticels, drying with marked ribs; seed 1, subturbinate, flattened, ca. 8.5 × 6 mm, laterally rugose.

Distribution: Colombia, Suriname, French Guiana(?), Ecuador, Peru, Brazil and Bolivia; over 35 collections studied, 3 from the Guianas (SU: 2; FG: 1).

Selected specimen: Suriname: langs Corantijn bij Wonotobo, BW 2853 (U).

Phenology: Flowering reported from September.

Note: In the Guianas *Cissus surinamensis* is known only from 2 collections from Suriname, the only specimen from French Guiana is sterile, probably a resprout on forest understory, and identified with doubt.

10. **Cissus tinctoria** Mart. in Spix & Martius, Reise Bras. 1: 368. 1823. Type: Brazil, without locality [probably Minas Gerais], "Tinta dos Gentios", Martius s.n. (holotype M).

Vitis selloana Baker in Mart., Fl. Bras. 14(2): 204. 1871. – *Cissus selloana* (Baker) Planch. in A. DC. & C. DC., Monogr. Phan. 5: 521. 1887. Type: Brazil, Minas Gerais, Lagoa Santa, Warming 1861 (holotype C).
Cissus boliviana Lombardi, Brittonia 50: 19. 1998. Type: Bolivia, La Paz, Prov. Nor Yungas, 2.5 km N (above) Yolosa on road to Coroico, Solomon & Kuijt 11619 (holotype LPB, isotype, MO), syn. nov.

Liana; drying blackish and brittle; trichomes unbranched and eglandular. Branches cylindric. Stipules triangular, ca. 5 × 3 mm, puberulent or glabrous, ciliate. Leaves simple, occasionally 3-5-lobed in vegetative branches; petiole 1.9-16 cm long; leaf blade membranous to chartaceous, slightly bullate, drying blackish, oblong in vegetative branches, oblong, triangular, or elliptic in reproductive branches, 5.1-16.3 × 3-20.4 cm, margin denticulate, apex acuminate, base cordate, subcordate, truncate, or cuneate, glabrous or puberulent along veins adaxially, puberulent abaxially. Inflorescence 4-9.2 × 3.3-6.8 cm; peduncle 1.9-4.4 cm long, puberulent, green; pedicels 2-3.5 mm long, glabrous, greenish. Flower buds ellipsoidal or conoidal (SE Brazil); perianth yellowish-green; calyx 1-2 × 1.5-3 mm, truncate, base truncate or rounded, glabrous; corolla in bud 1.5-2 × 1-2 mm, petals glabrous; anther dehiscence latrorse; disc yellowish. Berry purple, globose, ca. 8 mm, smooth; seed 1, subturbinate, rounded, ca. 6.5 × 4 mm, laterally slightly rugose.

Distribution: Suriname and Brazil; 65 collections studied, 2 from the Guianas (SU: 2).

Specimens examined: Suriname: Sipaliwini Distr., Central Suriname Nature Reserve, Rosário *et al.* 1885 and 2201 (HRCB, MO).

Phenology: Flowers and immature fruits in June.

Note: *Cissus tinctoria* has two disjunct populations, in SE Brazil and in N Brazil and Suriname. The leaves are smaller and the flower buds are ellipsoidal in specimens from the northern part of the distribution, while the ones from southeastern Brazil have bigger leaves and conoidal flower buds.

11. **Cissus trigona** Willd. ex Schult. & Schult.f., Mant. 3: 248. 1827. Type: Brazil, Pará, without locality, Hoffmannsegg s.n. (holotype B-W (not seen), isotype HAL).

Liana; glabrous or puberulent, trichomes unbranched, eglandular, rarely glandular. Branches 4-angled to 4-winged. Stipules deltoid, ca. 5 × 6 mm, gibbous, puberulent. Leaves trifoliolate; petiole 4.2-10.8 cm long; petiolule 0.2-2.5 cm long; leaflet blade chartaceous, elliptic, wide elliptic, or subelliptic, central one (9.4-)13.7-17.2(-23) × (3-)7.9-11.2(-14.4) cm, lateral ons (3.7-)10.1-11.3(-17.7) × (1.2-)5.1-7.3(-12.9) cm, margin denticulate, apex acute to acuminate, base cuneate, attenuate, or oblique, on both sides glabrous, sparse puberulent on veins on adaxial surface, or puberulent on abaxial surface. Inflorescence 4.5-9.3 × 2.8-5.4 cm; peduncle 2.3-5.6 cm long, puberulent at apex, green; pedicels ca. 4 mm long, glabrous or sparse glandular puberulent, greenish. Flower buds conoidal; perianth yellowish and pinkish; calyx 1 × 2 mm, truncate, base laterally expanded and discoid, truncate, glabrous; corolla in bud ca. 1.5 × 1 mm, petals glabrous, inside pinkish at apex, yellowish at base, caducous; anther dehiscence extrorse. Amphisarcum purple, globose, ca.

27 × 21 mm, with sparse lenticels; seed 1, subprismatic, flattened, ca. 15 × 8 mm, laterally smooth.

Distribution: Venezuela, Peru, Suriname, French Guiana, Brazil and Bolivia; over 40 collections studied, 4 from the Guianas (SU: 1; FG: 3).

Specimens examined: Suriname: 8 km in lijn van Pakapaka (Saramacca R.), naar de Ebbatop, Florschütz *et al.* 1607 (U). French Guiana: Inselbergs du Haut Marouini, de Granville *et al.* 16369 (CAY, HRCB); Mitaraka Sud, sommet Inselberg, Sarthou 907 (CAY, HRCB, P); Saül and vicinity, Boeuf-Mort trail, Mori *et al.* 23920 (CAY).

Phenology: Flowering reported from March, June, and November.

Note: *Cissus trigona* can be distinguished from *C. nobilis* by the angled or short winged branches (vs. conspicuously winged), the calyx base laterally expanded and discoid (vs. discoid or irregularly lobed), and by the amphisarcum fruits (vs. berries).

12. **Cissus ulmifolia** (Baker) Planch. in A. DC. & C. DC., Monogr. Phan. 5: 552. 1887. – *Vitis ulmifolia* Baker in Mart., Fl. Bras. 14(2): 213. 1871. Lectotype (designated by Lombardi 2000): Peru, Loreto, Maynas, Yurimaguas, Poeppig addenda 22 (hololectotype W).

Liana; glabrous or puberulent, trichomes unbranched, eglandular. Branches cylindric or 4-angled, sometimes with irregular wings. Stipules spathulate, 4 × 4 mm, gibbous, glabrous, ciliate. Leaves trifoliolate; petiole 1.2-17 cm long; petiolule 0-4.1 cm long; leaflet blade chartaceous, elliptic or subovate, the central one (3.6-)16.1-17(-22.7) × (1-)8.3-10.1(-14.8) cm, the lateral ones (1.9-)12.9-15.1(-18.7) × (0.7-)6.8-8.3(-12.3) cm, sometimes silvery on adaxial surface, margin denticulate, rare lobed, apex acute to acuminate, base rounded, attenuate, or cuneate, glabrous in both sides, or puberulent to sparse puberulent, mainly along the nerves on abaxial surface. Inflorescence 3.2-8.2 × 3.2-6.5 cm; peduncle 1.3-3.8 cm long, puberulent atapex to sparse puberulent, green; pedicels 2-4 mm long, puberulent, greenish or pinkish. Flower buds conoidal; perianth yellowish or pinkish; calyx 1 mm. long, 2 mm, lobes inconspicuous, urceolate or only slightly urceolate, base truncate to slightly lobed when dry, puberulent or glabrous; corolla in bud 2 × 1.5-2 mm, petals glabrous, caducous; anther dehiscence extrorse. Berry purple, fusiform or globose (C America), ca. 13 × 11 mm, with sparse lenticels, epicarp apparently though; seed 1, subbotuliform, rounded, ca. 12 × 9 mm, laterally markedly sulcate.

Distribution: Colombia, French Guiana, Ecuador, Peru, Brazil and Bolivia; over 35 collections studied, 1 from the Guianas (FG: 1).

Specimen examined: French Guiana: Piste de Risquetout, Prévost & Feuillet 3971 (CAY, U, MO).

Phenology: Flowering reported in September.

Note: *Cissus ulmifolia* is very close to *C. peruviana* Lombardi. The two species are partially sympatric in Colombia, Brazil (Acre), Ecuador, Peru and Bolivia, but *C. ulmifolia* differs from *C. peruviana* by the conoidal buds (vs. ellipsoidal), the lobed calyx base (vs. truncate) and by the leaves with a few conspicuous pearl glands on the abaxial surface (vs. many glands).

13. **Cissus venezuelensis** Steyerm., Bol. Soc. Venez. Ci. Nat. 26: 427. 1966. Type: Venezuela, Bolívar, between km 119 and 123 S of El Dorado, Steyermark *et al.* 92987 (holotype VEN (not seen)).

Liana; trichomes unbranched and eglandular. Branches cylindric. Stipules triangular, ca. 3.5 × 2 mm, gibbous, glabrous, ciliate. Leaves simple; petiole 1.3-1.6 cm long; leaf blade carnose, elliptic, 8.3-12.6 × 3-6.3 cm, margin inconspicuous denticulate, apex acuminate, caudate or mucronate, base cuneate, subdecurrent or rounded, glabrous or pulverulent on primary vein on adaxial surface, tertiary veins inconspicuous. Inflorescence sometimes falsely terminal, 4.6-5.7 × 3.2-3.9 cm; peduncle 2.7-3.5 cm long, sparse puberulent, green; pedicels 1.5-2 mm long, glabrous, greenish. Flower buds conoidal; perianth yellowish-green; calyx 1.5-2 × 2 mm, lobes triangular, base rounded, glabrous; corolla in bud 2-2.5 × 2 mm, petals glabrous; anther dehiscence latrorse; disc cruciform when dried. Mature berry and seeds not seen.

Distribution: Venezuela and Guyana; 4 collections studied, 2 from the Guianas (GU: 2).

Specimens examined: Guyana: Barabara Cr., Mazaruni R., Forest Dept. British Guiana F3361 (NY); Upper Mazaruni R., Ayanganna Mt., Tillett *et al.* 45890 (US).

Phenology: Flowering reported from May, immature fruits from August.

Notes: *Cissus venezuelensis* is restricted to the Venezuela-Guyana border area and rarely collected.

The falsely terminal inflorescence is formed by the abortion of subsequent nodes in the reproductive branches.

14. **Cissus verticillata** (L.) Nicolson & C.E. Jarvis, Taxon 33: 727. 1984. – *Viscum verticillatum* L., Sp. Pl. 1023. 1753. – *Phoradendron verticillatum* (L.) Druce, Rep. Bot. Soc. Exch. Club Brit. Isles 3: 422. 1914. Lectotype (designated by Nicolson & Jarvis 1984): Herb. Linnaeus 1166.10 (lectotype LINN (not seen)).

In the Guianas only: subsp. **verticillata**

Cissus sicyoides L., Syst. Nat. ed. 10: 897. 1759. – *Vitis sicyoides* (L.) Morales in Poey, Repert. Fis.-Nat. Isla Cuba 1: 206. 1865. – *Vitis vitiginea* (L.) W.L. Theob. var. *sicyoides* (L.) Kuntze, Revis. Gen. Pl. 1: 139. 1891. Lectotype (designated by Nicolson & Jarvis 1984): illustration in Browne, Civ. Nat. Hist. Jamaica 4, f. 1-2. 1756.
Cissus ovata Rich., Actes Soc. Hist. Nat. Paris 1: 106. 1792 (non Lam., Tabl. Encycl. 1: 331. 1792). – *Cissus obscura* DC., Prodr. 1: 629. 1824. Type: Brazil, Amapá, Oiapoque, Santalum, Leblond 77 (holotype G (not seen)).
Cissus puncticulosa Rich., Actes Soc. Hist. Nat. Paris 1: 106. 1792. Type: French Guiana, Cayenne, Leblond 79 (holotype G (not seen)).

Liana; trichomes unbranched and eglandular. New branches cylindric, old ones 4-angled. Stipules falcate or ovate, 2-8 × 1-2 mm, glabrous or pubescent, sometimes reflexed. Leaves simple; petiole (0.8-)3.4-7.8 cm long; blade chartaceous or carnose, ovate, triangular, elliptic or cordiform, (0.9-)3.5-22.7 × (0.8-)2.2-17.4 cm, margin denticulate or toothed, apex acute or acuminate, base cordate, truncate, cuneate or oblique, both sides glabrous or canescent, hispid on adaxial surface or villous on abaxial surface. Inflorescence 3.5-7.4 × 3-5.9 cm; peduncle (0.2-)1.9-5.2 cm long, glabrous or pubescent, green; pedicels 1-5.5 mm long, glabrous, greenish. Flower buds ellipsoidal; perianth yellowish; calyx 0.5-1 × 1-2.5 mm, truncate or rarely with rounded lobes, base rounded, glabrous; corolla in bud 0.5-2.5 × 1-2.5 mm, petals glabrous; anther dehiscence latrorse; disc yellowish. Berry purple, globose, 4-11 mm, smooth; seed 1, subturbinate, rounded, ca. 5 × 3 mm, laterally smooth.

Distribution: Southern U.S.A. to N Argentina and Uruguay, including Caribbean; over 2000 collections studied, 218 from the Guianas (GU: 42; SU: 104; FG: 72).

Selected specimens: Guyana: Koriabo, Barima R., Northwest Distr., van Andel *et al.* 1312 (BHCB, U); Pomeroon R., Pomeroon Distr., de la Cruz 3010 (F); Demerara-Mahaica Region, along Linden-Soesdyke Hwy., Pipoly 9247 (B, CAY, U, P, US); margins of Berbice R., S of New Dageraad, Maas *et al.* 5557 (F, P, S, U, Z); Upper Mazaruni R., de la Cruz 2375 (F); NW slopes of Kanuku Mts., drainage of Moku-Moku Cr., Smith 3506 (B, F, U, P, S, W). Suriname: along Paulus Cr., Lower Suriname R.,

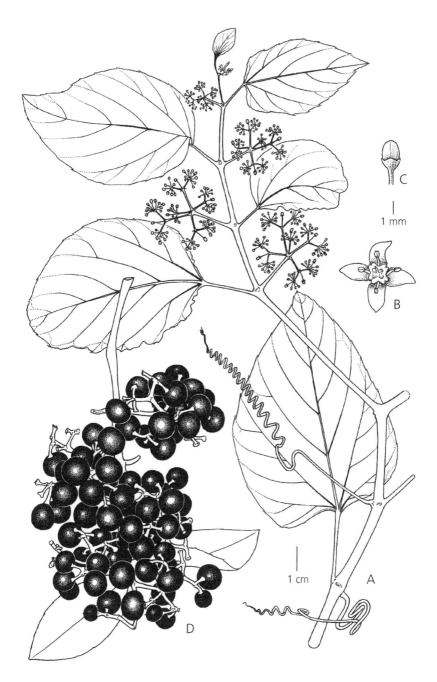

Fig. 5. *Cissus verticillata* (L.) Nicolson & C.E. Jarvis: A, habit; B, flower; C, flower bud; D, infructescence. Drawing by H. Rypkema.

Mennega 193 (BR, U); forest of Zandery, Samuels 504 (A); Paramaribo, Kramer & Hekking 2711 (U, Z); Saramacca, Everaarts 552 (U); Kabalebo Dam Project area, Lindeman & Görts-van Rijn *et al.* 101 (BHCB, CAY, U). French Guiana: Village Boni de Loca, Bassin du Maroni, Fleury 448 (BHCB, CAY); Ile de Cayenne, de Granville 5755 (BR, CAY, P); Point Diamante, Hahn 3523 (CAY, P, US); R. Mana, Cr. Arouany, Hallé 655 (P, US); Iles du Salut, Sagot 84 (BM, P, S, W); R. Inini, en amont de Maripasoula, Sastre *et al.* 4002 (NY, P).

Vernacular names: Guyana: bird-vine, snake-vine. Suriname: baaka-kifaia (Paramaccan, van Roosmalen, 1985), boenhatimama (Sranan), brakakijifaja (Matawai), kii faja (Ndjuka), paramaroe (Carib), tapoehoso wiwiri (Sranan). French Guiana: amatalea (Wayãpi), dontuwa (Boni), liane molle, pupuju (Wayana), quifaia (Taki-taki), weti kii faya (Boni).

Phenology: Flowering and fruiting reported along the year.

Note: *Cissus verticillata* subsp. *verticillata* is the only subspecies occurring in the Guianas and also the taxon with broadest geographic distribution in the genus. The other 3 subspecies have a chiefly Caribbean distribution: *C. verticillata* subsp. *colombiana* Lombardi occurs along the Caribbean sea shore in Panama, Colombia and Venezuela, and in the Virgin Islands, Leeward Islands, Windward Islands and southern Dutch Antilles; *C. verticillata* subsp. *micrantha* (Poir.) Lombardi occurs in Cuba, Haiti and Dominican Republic; and *C. verticillata* subsp. *oblongolanceolata* (Krug & Urb.) Lombardi in Cuba, Jamaica, Haiti and Dominican Republic.

131. MELIACEAE

by

TERENCE D. PENNINGTON[5] & NICOLA BIGGS[1]

Trees, treelets or rarely shrubs, monoecious, dioecious or polygamous. Buds protected by a cluster of scale-leaves or naked. Stipules absent. Leaves spirally arranged, usually pinnate, with or without a terminal leaflet, often with a terminal bud showing intermittent growth, less frequently 3-foliolate or 1-foliolate, rarely bipinnate; leaflets usually with entire margins, rarely serrate or crenate; venation usually eucamptodromous, less frequently brochidodromous; indument usually simple, less frequently of stellate, malpighiaceous or dibrachiate hairs, or peltate scales. Inflorescences usually axillary or in axils of scale leaves, less frequently ramiflorous or cauliflorous, usually paniculate with cymose branchlets (thyrsoid), less frequently racemose, spicate or fasciculate. Flowers bisexual or unisexual; rudiments of opposite sex well-developed in unisexual flowers; calyx usually shallowly or deeply 3-5(-7)-lobed, less frequently truncate or with free sepals, aestivation usually open, less frequently imbricate; petals 3-7, free or partially united, aestivation usually imbricate or valvate, less frequently quincuncial or contorted; filaments rarely completely free, usually partly or completely united to form a staminal tube, with or without appendages, staminal tube urceolate, cyathiform or cylindrical, margin entire or with small appendages alternating with anthers or rarely opposite them, anthers (4-)5-10(-14), hairy or glabrous, inserted apically on filaments or on margin of staminal tube, or within throat of staminal tube and partially or completely included; antherodes in female flowers smaller, slender, not dehiscing nor producing pollen; nectary (disk) intrastaminal, stipitate, annular, patelliform, cyathiform or absent, free from or partly fused to base of staminal tube or ovary; ovary superior, 2-13-locular, locules 1-, 2- or multi-ovulate, placentation axile, ovules collateral, superposed or biseriate, style-head capitate, conical, lobed or discoid; pistillode in male flowers smaller, usually with a more slender style, but often with well-developed abortive ovules. Fruits loculicidal, septicidal or septifragal capsules or rarely drupes; seeds either winged and then usually attached to a large woody columella, or unwinged and then usually with a fleshy arillode or sarcotesta, rarely with a corky or woody sarcotesta, usually without endosperm, occasionally endospermous; embryo with plano-convex or flat, collateral, superposed or rarely oblique cotyledons; radicle usually superior, less frequently abaxial, included, extending to surface or long exserted.

[5] Herbarium, Royal Botanic Gardens, Kew, Richmond, TW9 3AB, U.K.
Illustrations by Rosemary Wise.

Distribution: Pantropical, with ca. 700 species in ca. 50 genera; in the Guianas 42 species in 5 native genera, and 4 introduced species.

LITERATURE

Mabberley, D.J. 1984. A Monograph of Melia in Asia and the Pacific, the History of White Cedar and Persian Lilac. Gardens' Bulletin Singapore 37(1): 49-64.

Mabberley, D.J. & C.M. Pannell. 1995a. Meliaceae. In M.D. Dassanayake (ed.), Revised handbook of Flora of Ceylon 9: 229-300.

Mabberley, D.J., C.M. Pannell & A.M. Sing. 1995b. Meliaceae. In Foundation Flora Malesiana (eds.), Flora Malesiana 12(1): 1-407.

Palacios, W.A. 2007. Fam. 98 Meliaceae. In G. Harling & C. Persson (eds.), Flora of Ecuador 82: 1-90.

Pennington, T.D. 2006. Flora da Reserva Ducke, Brasil, Meliaceae. Rodriguesia 57(2): 209-246.

Pennington, T.D. & K.S. Edwards. 2001. Meliaceae. In J.A. Steyermark et al. (eds.), Flora of the Venezuelan Guayana 6: 528-549.

Pennington, T.D. & A.R.A. Görts-van Rijn. 1984. Meliaceae. In A.L. Stoffers & J.C. Lindeman (eds.), Flora of Suriname 5(1): 519-569.

Pennington, T.D. & B.T. Styles. 1975. A generic monograph of the Meliaceae. Blumea 22(3): 1-540.

Pennington, T.D. & B.T. Styles. 1981. Meliaceae. Flora Neotropica 28: 1-470.

Pennington, T.D. & B.T. Styles. 2001. Meliaceae. In W.D. Stevens et al. (eds.), Flora de Nicaragua 2: 1419-1430.

Styles, B.T. & F. White. 1991. Meliaceae. In R.M. Polhill (ed.), Flora of Tropical East Africa p. 1-68.

KEY TO THE GENERA
(including those introduced*)

1. Ovary with 1-2 ovules per locule; fruits loculicidal capsules or drupes. . . 2
   Ovary with 4-many ovules per locule, these biseriate; fruits septifragal or septicidal capsules 6

2. Leaflet margins serrate or crenate; fruits drupes. 3
   Leaflet margins entire; fruits capsules 4

3. Leaves 2-3-pinnate . . . . .7. Melia*
   Leaves pinnate . . . . .1. Azadirachta*

4   Anthers inserted at apex of filaments or on margin of staminal tube . . . .
. . . . 9. *Trichilia*
Anthers inserted within throat of staminal tube 5

5   Sepals 5, free, with quincuncial aestivation; nectary (disk) cyathiform. . . . .
. . . . .2. *Cabralea*
Sepals usually 4, partially united, with open aestivation; nectary (disk)
stipitate, often expanded to form a collar below ovary . . . . 5. *Guarea*

6   Stamens 5, filaments free, but adnate to androgynophore below, anthers 5,
inserted apically on filaments. . . . . .4. *Cedrela*
Staminal tube of completely united filaments, anthers 8-10, inserted within
throat of staminal tube; nectary (disk) annular or reduced to a short stipe
below ovary, never in the form of an androgynophore 7

7   Ovules 3-8 per locule; seeds large, angular, unwinged, surrounded by a thick
woody sarcotesta . . . . 3. *Carapa*
Ovules 12-16 per locule; seeds small, winged, without woody sarcotesta. . 8

8   Capsule globose, subglabrous or trigonous, not or scarcely longer than broad;
seeds orbicular to suborbicular, winged all the way round . . . . 6. *Khaya**
Capsule oblong-ovoid or oblong, elongate, at least 2 times as long as broad;
seeds with a single terminal wing. .8. *Swietenia**

1. **AZADIRACHTA** A. Juss., Bull. Sci. Nat. Géol. 23: 236. 1830.
Type: *A. indica* A. Juss.

Trees, polygamous. Bud-scales absent. Indument of simple hairs. Leaves
pinnate. Inflorescences axillary, many-flowered panicles. Hermaphroditic
and male flowers on same plant; calyx 5-lobed, lobes imbricate; petals 5, free,
imbricate; staminal tube cylindrical, slightly expanded at mouth, terminated
by 10 small appendages which are often united to form a frill, anthers 10,
inserted at base of and opposite appendages; nectary annular and fused to
base of ovary; ovary 3-locular, locules with 2 collateral ovules, style-head
expanded to form a ring bearing 3 acute, partially fused papillose stigmatic
lobes. Fruits 1(-2)-seeded drupes with a thin cartilaginous endocarp.

Distribution: Species 2, native to Indomalesia, of which *A. indica* is
introduced in tropical America.

1. **Azadirachta indica** A. Juss., Mém. Mus. Nat. Hist. Paris 19: 221,
t.13, f.5. 1830. Type: Sri Lanka, near Colombo, Hermann 161
(lectotype BM-HERM, designated by Mabberley 1995a).

Tree, to 16 m tall; bark red-brown or greyish, fissured and flaking in
older trees. Leaves imparipinnate or paripinnate with terminal spike, 4-7

pairs, red when young; leaflets opposite to subopposite; petiolule 1-2 mm; leaflet blade falcate-lanceolate, 5-9 × 1.5-3.5 cm, subglabrous, apex long acuminate, base asymmetric, acute, margin serrate. Inflorescence to 30 cm long, lax, paniculiform, axillary, sweetly scented; pedicels 2 mm, finely pubescent; calyx 1 mm, salveriform, lobes rounded, pubescent, margins ciliate; petals white, 4-6 mm long, linear spathulate, pubescent on both sides; staminal tube glabrous to sparsely pubescent, 10 ribbed, margin with 10 lobes, anthers 10, 0.8 mm long; ovary glabrous to finely pubescent. Fruit green becoming yellow when ripe, elliptic, 1-2 cm long, mesocarp thin, fleshy. (Mabberley, 1995a, 1995b).

Distribution: Introduced in tropical America; occasionally seen planted in parks and gardens; 1 collection studied (FG: 1).

Specimen examined: French Guiana: Kourou, Campus Silvolab, Barrabé 218 (CAY).

Notes: It is the only pinnate-leaved species of MELIACEAE with serrate leaflets in tropical America.

The terminal spike does not occur at the ends of all leaflets and sometimes there is a terminal leaflet instead.

2. **CABRALEA** A. Juss., Bull. Sci. Nat. Géol. 23: 237. 1830.
Type: *C. polytricha* A. Juss. [= C. canjerana (Vell.) Mart. subsp. polytricha (A. Juss.) T.D. Penn.]

Trees, dioecious (?). Bud-scales absent. Indument of simple hairs. Leaves pinnate with limited apical growth; leaflets glandular-punctate and -striate. Inflorescences in axillary panicles. Flowers unisexual (?); calyx of 5 free, quincuncial sepals; petals 5, free, imbricate or quincuncial; filaments completely united in a staminal tube terminated by short appendages alternating with anthers, anthers 10, glabrous, inserted within throat of staminal tube; nectary cyathiform; ovary 5-locular, locules with 2 superposed ovules, style-head discoid. Fruits (4)-5-valved loculicidal capsules, valves with 1-2 superposed seeds; seeds fleshy, partly surrounded by a free fleshy arillode.

Distribution: A single species with 2 subspecies throughout tropical America, from Costa Rica to northern Argentina.

1. **Cabralea canjerana** (Vell.) Mart., Syst. Mat. Med. Bras. 38. 1843.
– *Trichilia canjerana* Vell., Fl. Flum. [text] 176. 1829; [icones] 4: t. 102. 1831. Iconotype: in Vellozo, Fl. Flum. 4, t. 102. 1831.

In the Guianas only: subsp. **canjerana**

Tree, to 40 m tall, dioecious, larger specimens buttressed; bark pale grey fissured. Young branches appressed puberulous, lenticels sometimes present. Leaves paripinnate with limited apical growth, 30-70 cm long; petiole semiterete, rachis terete, appressed puberulous or glabrous; leaflets opposite, 7-12 pairs; petiolule (0.5-)1-4(-5) mm long; leaflet blade usually lanceolate, 10-16 × 5-6.5 cm, apex usually narrowly attenuate or acuminate, base strongly asymmetric, narrowly attenuate on one side and obtuse on the other, chartaceous, glabrous except for tufts of hairs in axils of secondary veins, moderately to densely glandular-punctate and -striate; venation eucamptodromous, midrib slightly prominent, secondaries 10-14 pairs, ascending, straight or slightly arcuate, parallel or slightly convergent, tertiaries obscure or absent. Inflorescence usually axillary, a laxly-branched panicle, 6-15 cm long, glabrous; pedicels 1-3 mm long. Flowers unisexual; calyx patelliform, sepals broadly ovate to suborbicular, 1-2 mm long, apex acute to rounded, sparsely to densely puberulous or pubescent, ciliate; petals usually quincuncial or imbricate, rarely contorted, oblong, 5-10 × 1.5-2.5 mm, apex rounded, glabrous; staminal tube cylindrical or less frequently narrowed at throat, 4.5-7 × 1.5-3(-4) mm, terminated by 10 short appendages alternating with anthers, glabrous, anthers 10, 0.8-1.2 mm long, glabrous; nectary 1-2 mm long, glabrous outside, short pubescent inside; ovary pubescent. Capsule globose or ellipsoid, 1.8-3 cm diam., with a rounded apex and tapering base, becoming wrinkled on drying and dark brown, with or without prominent pale lenticels, glabrous, 5-valved, valves with 1-2 superposed seeds, pericarp 1-5 mm thick, endocarp thin membraneous; seeds partially surrounded by a free fleshy arillode, seed coat thin, membraneous.

Distribution: Costa Rica and throughout tropical S America, Guyana, French Guiana; lowland rain forest on non-flooded sites; up to 600 m alt.; 2 collections studied (GU: 1; FG: 1).

Specimens examined: Guyana: N slope of Akarai Mts., drainage of Shodikar Cr. (Essequibo tributary), A.C. Smith 2927 (A, F, K, MO). French Guiana: Tumuc Humac Mts., NW extremity of crete du Miraraka, de Granville 1220 (CAY).

Phenology: Flowering from August to January throughout its range in S America, immature fruit from Guyana recorded in January.

Note: The other subsp. *polytricha* (A. Juss.) T.D. Penn. occurs in Brazil (Minas Gerais, Goiás).

68

3. **CARAPA** Aubl., Hist. Pl. Guiane 2 (suppl.): 32, tab. 387. 1775.
Type: *C. guianensis* Aubl.

Trees, monoecious. Shoot apex bearing a cluster of scale leaves. Indument of simple hairs. Leaves paripinnate with a dormant glandular leaflet at apex. Flowers unisexual, in large erect thyrsoid panicles, axillary or clustered at apex in axils of sterile bracts; calyx 4-5-lobed almost to base, lobes imbricate; petals 4-5, free, imbricate, spreading in open flowers; staminal tube cyathiform, urceolate or cylindrical, margin with entire or lobed appendages alternating with 8-10 anthers fixed within throat of tube; nectary well-developed, cushion-shaped, surrounding base of ovary and partially fused to it; ovary 4-5-locular, locules 3-8-ovulate, style-head discoid. Fruits pendulous sub-woody, subglobose, septifragal capsules opening by 4-5 valves from apex and base simultaneously; seeds large, angular, with a thick corky or woody sarcotesta.

Distribution: 3 species in tropical America, all occur in the Guianas.

LITERATURE

Kenfack, D., 2008. Systematics and evolution of Carapa (Meliaceae-Swietenoideae). Ph.D. Thesis. University of Missouri, St. Louis, U.S.A. 237 pp.
Scotti-Saintagne, C. *et al.* 2013. Phylogeography of a species complex of lowland Neotropical rain forest trees (Carapa, Meliaceae). Journal of Biogeography, 40: 575-592.

KEY TO THE SPECIES

1. Leaflets usually more or less elliptic with acute to acuminate apex; flowers usually sessile, predominantly 4-merous with 8 anthers, 4-locular ovary . . . . .2. *C. guianenesis*
   Leaflets generally oblong with rounded apex; flowers slender-pedicellate, predominantly 5-merous with 10 anthers, 5-locular ovary 2

2 Leaflet blade discolourous, tertiary venation loose and flat . . . . . 1. *C. akuri*
   Leaflet blade not discolourous, tertiary venation dense and raised . . . . 3. *C. surinamensis*

1. **Carapa akuri** Poncy, Forget & Kenfack, Brittonia 61(4): 367. 2009.
Type: Guyana, Upper Demerara-Berbice Reg., Mabura Hill, Forget 501 (holotype P, isotypes BRG, MO, US).

Buttressed canopy tree, to 35 m tall and 0.8-1 m dbh, glabrous, bole cylindrical; bark greyish and smooth on young individuals, flaking in rectangular patches in adult trees, slash reddish, exudating whitish-translucent sap. Young foliage yellow. Leaves paripinnate, crowded at end of branches, (40-)60-115 cm long; petiole 12-28 cm long, swollen at base, rachis (30-)43-90 cm long, glabrous; leaflets opposite, 6-13 pairs; petiolule 10-20 mm long; leaf blade discolorous, oblong, basal blade 9-20 × 5-10 cm, apical ones 16-56 × 4.5-13 cm, apex rounded to broadly acute, mucronate, glandular, base cuneate to rounded, slighly asymmetrical; midrib prominent on abaxial surface, secondary veins 8-20 pairs, tertiaries loose and flat. Inflorescence at end of branches in groups of 6-10, in axils of undeveloped leaves up to 3 cm long, a pendulous thyrse, (35-)60-100(-120) cm long, very much branched, branches transversely scurfy; pedicels (1.5-)2-3.5 mm long. Flowers with 1-3 in axil of 1 mm long bract, (4)5-merous; calyx green, lobes narrowly triangular to broadly ovate, 1-1.5 mm long, margin ciliolate; petals whitish to yellow-green, 5, free, oblong or obovate, 4-6 mm long; staminal tube white, urceolate, 3.5-5 mm long, margin with 10 truncate or more or less emarginate lobes, anthers or antherodes 10; nectary cushion-shaped, white; ovary ovoid to globose, 5-locular, locules with 4 ovules, style less than 0.7 mm long, stigma discoid, yellow; pistillode conical. Capsule green, becoming brown at maturity, globose to ovoid, 7-11 × 6-7 cm, valves with more or less developped warty excrescences and numeral extrafloral nectaries; seeds 2.5-4.8 × 3-5.5 cm, brown and smooth. First seedling leaves simple.

Distribution: Endemic to central Guyana; along large streams, in seasonally inundated forest and upland lateritic hills; 8 collections seen by authors of *C. akuri* (GU: 8).

Specimens cited: Guyana: Rupununi area, new road from Lethem to 25 km past Surama village, Acevedo-Rodriguez 3431 (BRG, CAY, US); Iwokrama Rainforest Reserve, Essequibo R., Kurupukari, N of Iwokrama base camp, Turtle Mt. transect, Mutchnik & Allicock 383 (BRG, CAY, US), Lady Smith Cr. Transect, Mutchnik 843 (BRG, US), Pisham Pisham transect, Clarke 365 (BRG, US), Akromuku transect at Akromuku F., Clarke 1304 (BRG, US), Malali Hill, Forget 576 (P); Tropenbos Pibiri Reserve, Forget 502 (BRG, P, US).

Vernacular name: Guyana: crabwood.

Phenology: Flowering during the dry season between November and February, fruiting in the rainy season between February and July.

Use: The straight bole produces good lumber that is used locally.

Notes: The authors of the MELIACEAE treatment could not study the material of this species. The description and the mentioned collections are taken from the original publication.

The most recently published study of the Carapa complex (Scotti-Saintagne *et al*. 2013), which entailed a comprehensive analysis of many genetic markers from throughout the Central American and South American range of the species, concludes that there are only two genetic clusters which support a taxonomic classification restricted to two species *C. guianensis* and *C. surinamensis*, as defined by Styles (1981), and there has been widespread gene flow and introgression events within them.

2. **Carapa guianensis** Aubl., Hist. Pl. Guiane 2 (suppl.): 32, tab. 387. 1775. Type: French Guiana, Aublet s.n. (holotype BM).

Buttressed tree, to 35 m tall and 1 m dbh; bark brown, scaling irregularly in plates, slash bright pink to red, fibrous, sapwood pinkish to cream. New foliage opening a characteristic wine-red colour. Young shoots massive, subglabrous, lenticellate. Leaves paripinnate, densely clustered at shoot apex, 50-90 cm long; petiole and rachis terete, glabrous; leaflets opposite, 5-9 pairs; petiolule 1-1.5 mm long; leaflet blade elliptic or oblong-elliptic, 18-25 × 6-9 cm, apex acute to acuminate, less frequently rounded, base acute to truncate, glabrous; venation mostly eucamptodromous, midrib flat or slightly raised adaxially, secondary veins 9-12 pairs, straight, parallel, intersecondaries short to moderate, tertiaries oblique, obscure. Inflorescence clustered around shoot apex above a cluster of scale-leaves, a lax-branched thyrse, 30-60 cm long, terminal cymules densely clustered, with scurfy pubescence; pedicels 0-2 mm long. Flowers scented, green below, white or creamy white and pink at apex; calyx 4-lobed to near base, lobes rounded, 1-2 mm long, imbricate, glabrous; petals 4, free, ovate to obovate, 4.5-6(-7) mm long, margin ciliolate; staminal tube urceolate, 3.5-4.5 mm long, margin with 8 truncate, rounded or variously lobed appendages, glabrous, anthers 8; nectary cushion-shaped; ovary glabrous, 4-locular, locules with 3-4 ovules; pistillode in male flowers slender, with vestigial ovules. Capsule globose or weekly quadrangular, 5-10 cm long, valves 4, obscurely ridged, glabrous, with a rough surface; seeds angular, 4-5 cm long, corky or woody. First leaves of seedling mostly 2-5-foliolate.

Distribution: C America, Greater and Lesser Antilles and northern S America; often on poorly drained and periodically flooded forest, but also present on non-flooded land; 42 collections studied (GU: 20; SU: 4; FG: 15).

Selected specimens: Guyana: Mabura region, Central Demerara compartment, main road to Pibiri, Ek *et al.* 624 (U). Suriname: Garnizoenspad, Baboenkreek, Saramacca R., BBS 2 (U). French Guiana: Oyapock, region de Camopi, between Saut Ouacarayou and Ilets Camopi, de Granville T-1093 (K); IRD, Prévost 4054 (CAY).

Vernacular names: Guyana: carapa, crabwood, karaba (Arawak). Suriname: krappa, krappaboom. French Guiana: carapa (Creole), carapa blanc, carapa rouge, karapa (Paramaka and Ndjuka), kaapa (Boni and Saramaka), karapa (Galibi), tiviru (Palikur), yani (Wayampi).

Phenology: Flowering from January to July, fruiting occurs throughout the whole year, with the capsules taking a year to mature.

Uses: It provides an important timber for construction and high class joinery. The seeds produce an oil which is widely used for lamps, soap, candle making and insect repellant. More recently it has become fashionable as an environmentally friendly skin-care lotion.

3. **Carapa surinamensis** Miq., Natuurk. Verh. Holl. Maatsch. Wetensch. Haarlem, ser. 2, 7: 75, tab. 19. 1851. – *Granatum surinamensis* (Miq.) Kuntze, Rev. Gen. Pl. 1: 110. 1891. Type: Suriname, Berg en Dal, Focke 1166 (holotype , isotype K).

Tree, to 30 m tall (often flowering when much smaller), unbuttressed, bole cylindrical with smooth pale bark. Young shoots massive, subglabrous. Leaves paripinnate, densely clustered at shoot apex, 40-90 cm long; petiole and rachis terete, glabrous; leaflets 5-8 pairs; petiolule 2-12 mm long; leaflet blade usually broadly oblong, 24-40 × 7-12 cm, apex rounded, glandular, base rounded or obtuse, glabrous; venation mostly eucamptodromous, midrib flat or raised adaxially, secondary veins 10-18 pairs, straight, parallel, tertiaries dense, and raised. Inflorescence axillary or subterminal in axils of scale-leaves, widely branched, 30-80 cm long, scurfy or glabrous; pedicels 2-5 mm long. Flowers scented, with pale cream corolla and staminal tube; calyx 1-1.5 mm long, 5-lobed almost to base, lobes rounded, imbricate, glabrous; petals 5, 4-8 mm long, glabrous, margin sometimes ciliate; staminal tube cyathiform or urceolate, 3-4.5 mm long, margin with 10 entire or lobed appendages, glabrous, anthers 10; nectary cushion-shaped, ribbed, glabrous; ovary glabrous, 5-locular, locules with 3-6 ovules; pistillode in male flower slender, with vestigial ovules. Capsule ovoid to subglobose, 7-10 cm long, valves 5, woody or leathery, with an obscure median ridge, glabrous, rough, lenticellate; seeds angular, 3-4 cm long, corky or woody. First seedling leaves simple.

Distribution: Guianas to central Amazonia; in lowland rain forest on both high (non-flooded) land and in gallery forest along rivers; 72 collections studied (GU: 7; SU: 40; FG: 25).

Selected specimens: Guyana: West Demerera, Mabura Hill, 180 km SSE of Georgetown, ter Steege *et al.* 578 (U). Suriname: Suriname R., Mapane region, Schulz in LBB 9317 (U). French Guiana: downstream from Base Camp near Piton rocheux remarquable, Cremers 5029 (U); Saül, Eaux Claires, Mori *et al.* 21528 (CAY).

Vernacular names: Guyana: crabwood. Suriname: kalappa, karaba, krapa, krappa. French Guiana: carapa (Creole), kalapa (Wayana).

Phenology: Flowering recorded from October to December, fruiting from Febuary to June.

Use: The timber is similar to that of *C. guianensis* and is marketed with it.

Note: Kenfack (2008) showed in his systematic study of the genus that it is better to use the name *C. surinamensis* instead of *C. procera* DC., which is an African species.

4. **CEDRELA** P. Browne, Civ. Nat. Hist. Jamaica 158, tab. 10, fig. 1. 1756.
Type: *C. odorata* L.

Trees, monoecious. Shoot apex bearing a cluster of scale leaves. Indument of simple hairs. Leaves usually paripinnate. Flowers unisexual, in large terminal, much branched thyrses; calyx lobed to near base, cup-shaped or shallowly toothed; petals 5, free, imbricate, adnate $1/3$ to $1/2$ their length to a columnar androgynophore (nectary) by a median keel; stamens 5, free but adnate to androgynophore below; ovary 5-locular, borne at apex of androgynophore, locules with 6-12 ovules, style-head discoid. Fruits woody, septicidal capsules, opening from apex by 5 valves, with a central woody columella; seeds dry, with a terminal wing, attached by seed end to apex of columella and winged towards base of capsule.

Distribution: About 17 species confined to the Neotropics, distributed from Mexico to northern Argentina; 2 species in the Guianas.

KEY TO THE SPECIES

1. Leaflets 8-16 pairs, pubescent to tomentose below; terminal cymules of inflorescence densely-flowered; petals usually pinkish; capsule 7-11 cm long . . . . 1. *C. fissilis*
Leaflets 5-10 pairs, glabrous; terminal cymules of inflorescence open, lax-flowered; petals greenish-white; capsule 2-5 cm long . . . . . . 2. *C. odorata*

1. **Cedrela fissilis** Vell., Fl. Flum. [text] 72. 1825; [icones], t. 68, f. 2. 1835. Lectotype: Plate in Vellozo, Fl. Flum. t. 68, 1835.

Tree, to 40 m tall, more than 1 m at dbh; bark greyish-brown, deeply fissured, with ridges scaling, slash pink, fibrous. Young shoots pubescent to subglabrous, smooth, with pale lenticels. Leaves paripinnate, 25-70 cm long; petiole and rachis semiterete, pubescent to tomentose at first, becoming glabrous; leaflets 8-16 pairs, opposite; petiolule 1-1.5 mm long; leaflet blade lanceolate, 8-14 × 2.5-4 cm, apex narrowly acuminate, base asymmetrical, obtuse, rounded or truncate, pubescent to tomentose abaxially; venation eucamptodromous, midrib slightly raised adaxially, secondary veins 14-17 pairs, slightly arcuate and convergent, intersecondaries short to moderate, tertiaries reticulate. Inflorescence widely branched, 50-90 cm long, ultimate cymules usually congested; pedicels 1-2 mm long; calyx cyathiform, shallowly lobed, 1.5-2.5 mm long, densely pubescent; petals often tinged pink, 8-10 mm long, densely tomentose on both sides; filaments 1.5-2.5 mm long, glabrous, anthers ca. 1.5 mm long; ovary ovoid to globose, 5-locular, locules 8-12-ovulate; pistillode in male flower slender, with vestigial ovules. Capsule obovoid, 7-11 cm long, pendulous, 5-valved, valves woody, dark brown with dense pale lenticels, glabrous, columella with 5 prominent broad wings; seeds 2.5-4.5 cm long (including wing), dark brown.

Distribution: Ranging from Costa Rica to SE Brazil and Bolivia; lowland forest from sea level to ca. 800 m alt.; 1 collection studied (GU:1).

Specimen examined: Guyana: Kanuku Mts., Wabuwak, Forest Dept. 5790 (= WB 377) (K).

Phenology: In Guyana collected with immature fruits in October.

2. **Cedrela odorata** L., Syst. Pl. ed. 10, 940. 1759. Type: Plate 10, fig. 1 of P. Browne, Civ. Nat. Hist. Jamaica 158. 1756.

Deciduous tree, to 40 m tall, 1.5 m dbh, with small buttresses, bole cylindrical; bark greyish brown fissured, ridges scaling; slash pink, fibrous, bitter; crushed twigs, leaves and fruit often smelling of garlic. Young shoots usually glabrous, with conspicuous lenticels. Leaves paripinnate, 25-70 cm long; petiole semiterete, rachis terete, puberulous at first; leaflets 5-10 pairs, usually opposite; petiolule 1-10 mm long; leaflet blade ovate, oblong-lanceolate or lanceolate, 6.5 -12 × 2.5- 4 cm, usually somewhat asymmetrical, apex acute to acuminate, base asymmetrical, acute to rounded, usually glabrous, with some hollow domatial cavities in axils of secondary veins abaxially; venation eucamptodromous, midrib flat or slightly raised adaxially, secondaries 8-12 pairs, slightly arcuate and convergent, higher order venation reticulate. Inflorescence a widely branched panicle, 15-50 cm long, cymules lax, puberulous; pedicels 1-2 mm long; calyx cyathiform or broadly tubular, 2-3 mm long, puberulous or glabrous; petals pale greenish-cream, 7-8 mm long, pubescent on both sides; filaments 2-3 mm long, glabrous, anthers ca. 1.5 mm long; ovary ovoid, glabrous, 5-locular, locules with 10-14 ovules; pistillode in male flowers slender, with vestigial ovules. Capsule ellipsoid to obovoid, 2-5 cm long, pendulous, 5-valved, valves thinly woody, grey-brown or brown with prominent pale lenticels, columella with 5 broad wings; seeds 2-3 cm long (including wing), light brown.

Distribution: Ranging from Mexico and C America, tropical S America to Paraguay and Argentina; occurs in both dry and moist lowland deciduous forest; 20 collections studied (GU: 7; SU: 9; FG: 4).

Selected specimens: Guyana: Cuyuni-Mazaruni R., Swarima Isl., Cuyuni R., Forest Dept. 2326 (= D 333) (K, NY). Suriname: Brokopondo, Sectie O, BW 3957 (NY). French Guiana: Pic Matecho, 22.5 km NE of Eaux Claires, Mori & Smith 25060 (NY); Monts d'Arawa, Sabatier & Molino 5019 (CAY).

Vernacular names: Guyana: red cedar. Suriname: ceder.

Phenology: Flowering in the dry season (August to September), fruiting from April to May.

5. **GUAREA** F. Allam. ex L., Mant. Pl. 150, 228. 1771.
    Type: *G. trichilioides* L. [= G. guidonia (L.) Sleumer]

Trees or treelets, dioecious. Bud-scales absent. Indument of simple hairs. Leaves nearly always pinnate with a terminal bud showing intermittent growth; leaflets sometimes glandular-punctate and glandular-striate. Inflorescences axillary, ramiflorous and cauliflorous, panicles, racemes or spikes. Flowers unisexual; calyx cup-shaped with an entire margin to deeply 3-7-lobed, aestivation open; petals 4-6(-7), free, nearly always valvate; staminal tube cylindrical, sometimes contracted at throat, margin entire, crenate or with small lobes, anthers 8-12, glabrous, inserted within throat of tube, completely included or partly exserted, alternated with lobes; nectary stipitate, nearly always expanded at apex to form a collar below ovary; ovary 2-10-locular, locules with 1-2 superposed ovules, style-head discoid. Fruits 2-10-valved loculicidal capsules, valves 1-2-seeded, pericarp leathery or woody, endocarp thin and cartilagineous; seeds with a thin, fleshy sarcotesta; embryo with plano-convex, nearly always superposed cotyledons, radicle abaxial, extending to surface; endosperm absent.

Distribution: About 65 species in the Neotropics and 5 in tropical Africa; 16 species in the Guianas.

KEY TO THE SPECIES

1. Petals generally 4-12 mm long; ovary (2-)4-6-locular, locules 1-ovulate; capsules (2-)4-6-valved, valves 1-seeded. 2
Petals generally 10-20 mm long; ovary 2-12-locular, locules with 2 superposed ovules; capsules 4-12-valved, valves with 1-2 superposed seeds 18

2. Capsules ≥ 3.5 cm long. 3
Capsules usually 1-3 cm long, never more than 3.5 cm long. 6

3. Leaves without a terminal bud; ovary 2-3-locular, glabrous; capsule constricted between seeds . . . . 14. *G. silvatica*
Leaves with a terminal bud; ovary (3-)4-6-locular, pubescent or strigose; capsules not constricted between seeds 4

4. Secondary veins 7-10 pairs; petals 6.5-7.5 mm long . . . . . 3. *G. convergens*
Secondary veins 12-18 pairs; petals 8-14 mm long 5

5. Inflorescence 3-12 cm long; petals 11.5-14 mm long; capsule 5-6.2 cm long. . . . . .2. *G. cinnamomea*
Inflorescence 15-40 cm long; petals 8-12 mm long, capsule 2.5-3.2 cm long. . . . . .6. *G. gomma*

6. Leaflets with tuft of hairs (domatia) at axils of secondary veins (lower surface). . . . . .5. *G. glabra*
Leaflets without domatia. 7

7. Capsules smooth 8
Capsules ribbed, winged or tuberculate 15

8. Capsules tomentose, pubescent or puberulous 9
Capsules glabrous. 14

9. Leaf blade pubescent or puberulous on abaxial surface 10
Leaf blade glabrous on abaxial surface. 11

10. Leaf tertiary veins not finely parallel and oblique; ovary tapering to apex, appressed puberulous; capsule pyriform to globose, minutely puberulous . . . . .10. *G. macrophylla*
Leaf tertiary veins numerous, fine, parallel, oblique; ovary rounded at apex, densely long-strigose; capsule depressed-globose to globose or pyriform, usually coarsely pubescent. . . . . .12. *G. pubescens*

11. Leaf tertiary veins numerous, fine, parallel, oblique; ovary rounded at apex, densely long-strigose; capsule depressed-globose to globose or pyriform, usually coarsely pubescent.........................*12. G. pubescens*
Leaf tertiary veins not finely parallel and oblique, often obscure; ovary tapering to apex, pubescent or strigose; capsules pyriform to globose, minutely puberulous or shortly pubescent 12

12. Leaf without indeterminate apical growth (without apical bud); secondary veins mostly 12-18 pairs; petals 4-6; anthers 8-11..........*6. G. gomma*
Leaves with indeterminate apical growth (with apical bud); secondary veins mostly 7-14 pairs; petals 4-5; anthers commonly 8 13

13. Inflorescence little-branched, pubescent or puberulous .*10. G. macrophylla*
Inflorescence a widely branched panicle with long weak slender axes, subglabrous.. *13. G. scabra*

14. Leaf with tufts of hairs in axils of secondary veins below; capsule pink or red, usually not lenticellate.......................................
....*5. G. glabra*
Leaf without tufts of hairs in axils of veins below; capsule dark brown, shining, usually with conspicuous pale lenticels ....*8. G. guidonia*

15. Young shoot velutinous or tomentose; capsule with 12-16 thin, convoluted and anastomosing longitudinal wings of 5-7 mm deep ... *16. G. velutina*
Young shoots not velutinous or tomentose; capsules verrucose, ribbed or shallowly winged, ribs sometimes torulose or tuberculate........... 16

16. Capsules contracted at base into a stipe of 0.5-0.8 cm long... *4. G. costata*
Capsules without a stipe, or if present then only 0.1-0.2 cm long 17

17. Leaf tertiary veins numerous, fine, parallel, oblique; ovary rounded at apex, densely long-strigose; capsule depressed globose to globose or pyriform, usually coarsely pubescent.........................*12. G. pubescens*
Leaf tertiary veins not finely parallel and oblique; ovary tapering to apex, glabrous, pubescent or strigose; capsules obovoid, pyriform, globose or rarely depressed globose, mostly puberulous.................... 18

18. Ovary glabrous 19
Ovary pubescent 20

19. Leaves with a terminal bud; ovary (3-)4-locular; capsule ellipsoid or globose or irregularly shaped, usually with large pale lenticels ...*9. G. kunthiana*
Leaves without a terminal bud; ovary 2-3-locular; capsule constricted between seeds, without lenticels .... 14. *G. silvatica*

20. Indument of young parts and on abaxial surface of leaf blade prominent, tomentose, coarsely pubescent or uniformly pale crisped puberulous.. 21
Indument of young parts fine, minute, puberulous, abaxial Surface of leaf blade sparsely and minutely puberulous or glabrous 26

21. Capsules winged or ribbed 22
Capsules smooth.. 24

22. Leaves bullate; mature capsule with thin fleshy convoluted longitudinal wings .... 11. *G. michel-moddei*
    Leaves (not bullate); mature capsules ribbed or tuberculate 23

23. Indument on leaf abaxial surface conspicuous, coarsely pubescent; secondary veins 12-17 pairs; petals 14-17 mm long; ovary 5-6-locular; capsule 3-4.8 cm long .... 1. *G. carinata*
    Indument on leaf abaxial surface inconspicuous, shortly and finely pubescent, secondary veins mostly 7-10 pairs; petals 6-9 mm long; ovary usually 4-locular; capsule 1.5-2.5 cm long .... 10. *G. macrophylla*

24. Petals 6-9 mm long .... 10. *G. macrophylla*
    Petals 10-20 mm long 25

25. Leaflet blade abaxially puberulous or tomentose, confined to midrib and veins .... 7. *G. grandifolia*
    Leaflet blade abaxially uniformly puberulous, with crisped pale whitish hairs .... 15. *G. trunciflora*

26. Capsules tuberculate .... 10. *G. macrophylla*
    Capsules smooth 27

27. Capsules 5-6 cm long .... 7. *G. grandifolia*
    Capsules 1.5-3.5 cm long 28

28. Capsules less than 2.5 cm long .... 10. *G. macrophylla*
    Capsules 2.5-3.2 cm long .... 6. *G. gomma*

1. **Guarea carinata** Ducke, Trop. Woods 76: 16. 1943. Type: Brazil, Amazonas, Esperança, mouth of R. Javarí, Ducke 1060, (holotype RB, isotypes K, MG, NY, US).

Tree, to 20 m tall, dioecious; bark soft, brown, suberous, scaling in thin irregular patches, longitudinally-fissured, 1-2 mm thick, twigs and branches also become strongly suberized and the corky bark is often inhabited by small ants. Young branches very stout, golden-tomentose at first, soon glabrous. Leaves pinnate, with a terminal bud showing intermittent growth; petiole 5-10 cm long, semiterete, rachis 3-60 cm long, terete, both tomentose at first, becoming glabrous and suberous; leaflets opposite, 2-10 pairs, lowest ones often much shorter and broader; petiolule 2-6(-8) mm long; leaflet blade chartaceous, usually oblong, less frequently elliptic or oblanceolate, 12-27 × 5.8-9(-12) cm, apex attenuate, obtuse or truncate, base obtuse to truncate, adaxial surface glabrous except for tomentose to pubescent midrib, lower midrib densely pubescent with rather coarse crisped hairs, sparser abaxially, sometimes obscurely glandular-punctate and -striate; venation eucamptodromous or brochidodromous, midrib sunken adaxially, secondaries 12-17 pairs, ascending, straight, parallel, intersecondaries short, tertiaries moderately prominent, oblique, parallel. Inflorescence axillary, 2-10.5 cm long, lower branches to 2.5 cm long or unbranched, flowers in small sessile clusters subtended by bracts, rather densely-flowered, tomentose; pedicels 1-2 mm long or flowers sessile. Calyx green or red, cyathiform, 3-7 mm long, 4-lobed, lobes triangular or ovate, 1-3 mm long, densely pubescent to tomentose outside; petals cream, 4, valvate, strap-shaped, 14-17 × 2-4 mm, apex acute, densely golden strigose outside, glabrous inside; staminal tube cream, 11-13 × 3-4 mm, margin undulate or crenulate, glabrous, anthers 8-11, 1.5-1.8(-2) mm long; antherodes similar but narrower, not dehiscing, without pollen; nectary with a stout stipe expanded to form an annulus below ovary, 1.5-2 mm long, glabrous; ovary densely strigose, (4-)5-6-locular, locules with 2 superposed ovules, style glabrous above; pistillode similar, with well-developed abortive ovules. Capsule purplish-red, depressed globose to pyriform, 3-4.8 × 3-4.2 cm, sometimes apiculate, apex truncate, base rounded or tapering, 5-6(-9)-valved, valves with 2 superposed seeds, valves shallowly to prominently 3-ribbed, central rib longer than and often branched above and anastomosing with lateral ones, tomentose when young and pubescent at maturity, pericarp ca. 5 mm thick; seeds ca. 1.5 × 0.9 cm, surrounded by a thin orange sarcotesta.

Distribution: The Guianas, Amazonia Brazil, Peru and Ecuador; a tree of non-flooded lowland forest; 4 collections studied (GU: 1; SU: 3).

Specimens examined: Guyana: Bartica-Potaro, 107 miles, Potaro R., Mahdia, Forest Dept. 3807 (= F 1071) (K). Suriname: Sipaliwini savanna, Oldenburger *et al.* 1111 (U), 1254 (U). French Guiana: Camp Caiman, route de Kaw, Grenand 1527 (CAY).

Phenology: Flowers in July and August and its fruits are ripe from October to January.

Note: The most striking feature of this species is the large purplish-red carinate fruit which is borne in clusters of 2-6, usually on the leafless part of the twigs and branches. The thick pericarp is tough but fleshy.

2. **Guarea cinnamomea** Harms, Notizbl. Bot. Gart. Berlin-Dahlem 13: 504. 1937. Type: Brazil, Amazonas, Paranagua, Krukoff 4556 (holotype B, destroyed, lectotype K, designated here, isolectotypes A, G, NY, U). – Fig. 1

Tree, to 20 m tall; bark brown, soft and rather suberous, usually scaling in thin irregular pieces. Young branches stout, puberulous at first, becoming glabrous. Leaves pinnate with terminal bud showing intermittent growth,

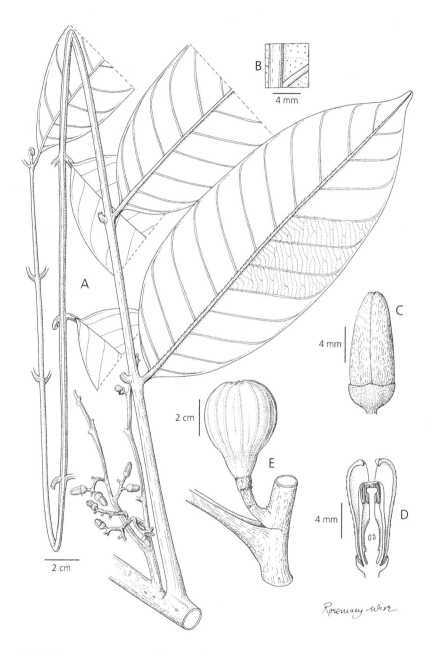

Fig. 1. *Guarea cinnamomea* Harms : A, habit; B, detail of leaflet abaxial surface; C, flower bud; D, flower, longitudinal section; E, fruit. (A-D, Daly *et al.* 4179; E, Thomas *et al.* 5362).

15-80 cm long; petiole semiterete when young, puberulous or glabrous; leaflets 2-11 pairs; petiolule 5-10 mm long; leaflet blade coriaceous, oblong, less frequently ovate, 11-17 × 4-7 cm, apex acuminate or occasionally acute, base usually broadly cuneate, adaxial surface minutely puberulous on midrib and veins, abaxial surface glabrous, with numerous red papillae, not glandular-punctate or -striate; venation eucamptodromous, midrib sunken adaxially, secondaries 12-17 pairs, parallel, straight for greater part of length, intersecondaries absent, tertiaries oblique, parallel. Inflorescence axillary, thyrsoid or racemose, 3-12 cm long, few-flowered, lower branches to 6 cm long, puberulous; pedicels broad, 1-1.5 mm long. Flower-buds acute, tapering strongly from near base to apex; calyx cyathiform, 2.5-3 mm long, margin truncate or irregularly and shallowly 4-5-toothed, minutely puberulous outside; petals cream coloured, 4-5, valvate, strap-shaped, 11.5-14 × 2-3 mm, acute, densely appressed pubescent outside, glabrous inside; staminal tube cylindrical, 3 × 3 mm, margin undulate, glabrous, anthers 7-9, 1.5-2 mm long; nectary with a stout stipe expanded into an annulus below ovary, 2 mm long, glabrous; ovary densely appressed pubescent to sericeous, 4-6-locular, locules 1-ovulate, style glabrous. Capsule ovoid or pyriform, 5-6.2 × 3.6-4.3 cm, sometimes apiculate, tapering gradually to a short stout stipe, smooth or faintly ribbed (5-8 ribs on each valve), puberulous or densely papillose, 4-6-valved, valves 1-seeded, pericarp 4-7 mm thick; seeds 3-6 developing in each capsule, 2.3-2.6 × 1.1-1.6 cm, surrounded by a thin fleshy vascularised sarcotesta, seed coat thin, soft.

Distribution: French Guiana and central and western Amazon basin; in lowland, non-flooded forest; 2 collections studied (FG: 1).

Specimen examined: French Guiana, Approuague R., R. Matarony, Saut Magasin, Oldeman B-1017 (CAY, K, U).

Phenology: Flowering in May and June and fruiting from August to April.

3. **Guarea convergens** T.D. Penn., Fl. Neotrop. 28: 260. 1981. Type: Brazil, Amazonas, Manaus to Caracarai road, km 154, Pennington *et al.*, 9967 (holotype INPA, isotypes FHO, MO, NY).

Tree, to 15 m tall, bole cylindrical; bark brown, scaling. Young shoots appressed puberulous at first, soon glabrous, greyish brown, smooth. Leaves pinnate, up to 50 cm long, with a terminal bud showing intermittent growth; petiole 6-8 cm long, semiterete, glabrous, rachis up to 25 cm long, terete, glabrous; leaflets 2-4 pairs; petiolule 2-4 mm long; leaflet blade coriaceous, elliptic, 7-13 × 3.5-5 cm, apex acuminate, base acute to cuneate, glabrous, not glandular-punctate or -striate; venation eucamptodromous, midrib sunken adaxially, secondary veins 7-10 pairs, arcuate-ascending, convergent, tertiary venation obscure, reticulate to oblique. Inflorescence a axillary, slender pyramidal thyrse, 5-11 cm long, puberulous; pedicels 0.5-1.5 mm long. Flowers with calyx pinkish-purple, patelliform, 1-1.5 mm long, irregularly lobed, sparsely puberulous outside; petals cream-coloured, 4, valvate, oblong or lanceolate, 6.5-7.5 × 1.5-2.5 mm, apex acute, appressed puberulous outside, glabrous inside; staminal tube 5-5.5 × 1.5-3 mm, margin undulate, glabrous, anthers 7-8, 0.6-0.8 mm long; nectary with a stipe broadened into a collar below ovary, glabrous; ovary strigose, 4-locular, locules 1-ovulate, style glabrous. Capsule reddish, globose, 3.5-4.5 cm long, apex truncate, base slightly attenuate, minutely papillose, 4-valved, valves obscurely 6-7-ribbed, 1-seeded, pericarp 4-5 mm thick; seeds ca. 2 cm long, surrounded by a thin fleshy sarcotesta.

Distribution: S Venezuela, French Guiana, central and western Brazilian Amazonia and Amazonian Peru; in lowland rain forest on non-flooded land; 8 collections studied (FG: 1).

Specimen examined: French Guiana: Bassin du Sinnamary, Crique Plomb, Loubry 1774 (K).

Phenology: Flowering in August.

Note: See note to *G. scabra* for differences with this species.

4. **Guarea costata** A. Juss., Bull. Sci. Nat. Géol. 23: 240. 1830. Type: French Guiana, Cayenne, Perrottet s.n. (holotype P).   – Fig. 2

Tree, 4-18 m tall; bark mid-brown. Young branches appressed puberulous becoming glabrous. Leaves pinnate with a terminal bud showing intermittent growth, to 40 cm long; petiole 4-11 cm long, semiterete, rachis 17-35 cm long, terete or canaliculate above, subglabrous; leaflets up to 6 pairs; petiolule 1-5 mm long; leaflet blade subcoriaceous, usually oblanceolate, less frequently broadly oblong, 13-20 × 5-6.6 cm, apex narrowly acuminate to short and obtusely cuspidate, base narrowly attenuate, glabrous, not glandular-punctate or -striate; venation eucamptodromous, midrib slightly sunken adaxially, secondaries 8-11 pairs, ascending, arcuate, slightly convergent, intersecondaries and tertiaries usually obscure. Inflorescence axillary or several on a short axillary shoot, indeterminate with intermittent growth from an apical bud, slender thyrse, 6-12 cm long, with flowers arranged in sessile 3-flowered cymules along main axis or branched with a few wide-spreading lateral branches to 7 cm long, sparsely appressed puberulous; pedicels ca. 0.5 mm long. Flowers with calyx cyathiform, 1.5-2 mm long, margin irregularly and often deeply divided into 3-4 broadly ovate, acute to obtuse lobes, sparsely appressed puberulous; petals 4, valvate, strap-shaped, 5.5-6.5 × ca. 1.5 mm, apex acute, minutely appressed puberulous or strigulose outside, glabrous inside; staminal tube ca. 5 × ca. 1.5 mm, margin undulate or truncate, glabrous, anthers 8, ca. 0.8 mm long; nectary with a stipe expanded to form an annulus beneath ovary, 1-1.5 mm long, glabrous; ovary densely stiff pubescent, 4-locular, locules 1-ovulate, style stout tapering from base to apex, glabrous above. Capsule brownish

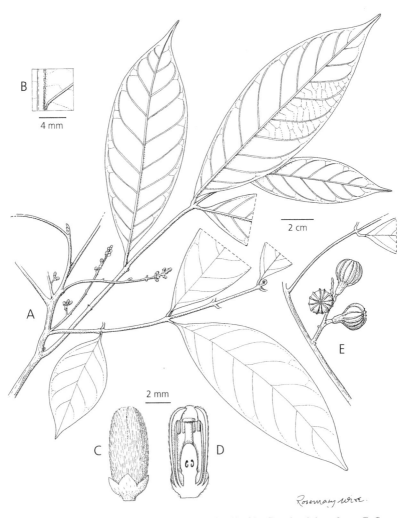

Fig. 2. *Guarea costata* A. Juss.: A, habit; B, detail of leaflet abaxial surface; C, flower bud; D, flower, longitudinal section; E, infructescence. (A-D, Lindeman *et al.* 130; E, Lanjouw & Lindeman 2500).

green, globose or depressed globose, 1.5-2 cm diam., contracted suddenly at base into a stipe 0.5-0.8 cm long, 4-valved, valves 1-seeded, each valve with 3 ribs or narrow wings to 2.5 mm deep, one rib or wing often more prominent than others, ribs sometimes branched and anastomosing, minutely appressed puberulous, hairs intermixed with red papillae, pericarp ca. 2 mm thick; seeds 1-1.2 × 0.5-0.8 cm, shaped like segment of an orange, surrounded by a thin orange sarcotesta.

Distribution: Suriname and French Guiana; a tree of non-flooded tropical rain forest; 16 collections studied (SU: 12; FG: 3).

Selected specimens: Suriname: Nassau Mts., Lanjouw *et al.* 2500 (K, U); Lely Mts., Lindeman & Stoffers *et al.* 236 (K); Brokopondo district, ESE of village Brownsweg, van Donselaar 2275 (U). French Guiana: Mt. Bellevue de l'Inini, de Granville *et al.* 7652 (CAY, K, U), 8072 (K, U); Mont Atachi Bacca, region de l'Inini, de Granville *et al.* 10707 (K).

Vernacular name: Suriname: doifi-siri (Sranang).

Phenology: Flowering is recorded from July to September and fruiting in March.

Note: *Guarea costata* is allied to several of the small-flowered *Guarea* species which have a 4-locular ovary with 1-ovulate locules, such as *G. scabra* and *G. velutina*. It differs from them both in its stipitate capsule and from *G. scabra* in its much smaller flowers and in the structure of the inflorescence. *G. velutina* also differs in the indument of the leaflets and in the much larger and more prominently winged capsules.

5. **Guarea glabra** Vahl, Eclog. Am. 8. 1807. Type: Lesser Antilles, St. Croix, Montserrat, Ryan s.n. (holotype C).

    In the Guianas only: subsp. **glabra**   – Fig. 3

    Guarea schomburgkii C. DC., Monogr. Phan. 1: 565. 1878. Type: Guyana, Ro. Schomburgk ser. II, 779 (= Ri. Schomburgk 1434) (lectotype G-DC, designated by Pennington 1981, isolectotypes G, K, NY, P).

Tree, to 25 m tall (often flowering when less than 10 m), dioecious, larger specimens usually have small buttresses; bark shallowly fissured or scaling, usually pale greyish or greyish-brown. Young branches sparsely pubescent, soon becoming glabrous, pale greyish-white, with scattered lenticels. Leaves pinnate, 5-30 cm long, terminal bud showing intermittent growth; petiole and rachis terete, at first minutely appressed pubescent, soon becoming glabrous; leaflets opposite, 2-4 pairs; petiolule 1-6(-10) mm long; leaflet blade chartaceous, elliptic to oblanceolate, 10-21 × 4.5-7.5 cm, apex acuminate, base usually attenuate or acute,

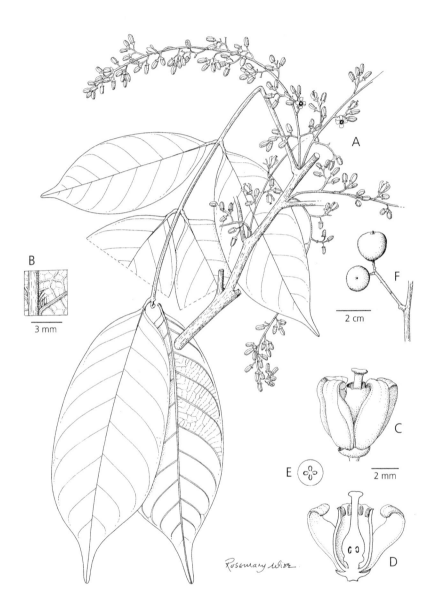

Fig. 3. *Guarea glabra* Vahl : A, habit; B, detail of leaflet; C, flower; D, flower, longitudinal section; E, ovary section; F, fruit. (A-E, Roberts 16283; F, BW 6677).

glabrous on both sides except for a tuft of hairs in axils of secondary veins abaxially, usually sparsely glandular-punctate and -striate; venation eucamptodromous, midrib flat or slightly sunken adaxially, secondaries 8-10 pairs, ascending, strongly arcuate and convergent, intersecondaries absent, tertiaries oblique, parallel. Inflorescence axillary, panicle, 10-15 cm long, with short lateral branches 1-5 cm long, ultimate branchlets often cymose or racemose, sparsely appressed puberulous; pedicels 1-3 mm long. Calyx patelliform, 0.5-1 mm long, with 4 shallow acute teeth, appressed puberulous; petals 4(-5), valvate, strap-shaped, 5-6 × 1-1.5 mm, apex acute, sparsely appressed pubescent or glabrous outside, glabrous within; staminal tube 4-6 × 1.5-2 mm, margin undulate, glabrous, anthers (7-)8; antherodes smaller, indehiscent, without pollen; nectary in male flowers stout- to slender-stipitate, expanded below ovary to form a collar, 0.5-1.5 mm long, glabrous, in female flowers shorter and reduced to an obscure swelling beneath ovary; ovary glabrous, 4-locular, locules 1-ovulate, style stout, glabrous. Capsule pink or red, globose, 1-1.5 cm long, usually slightly broader than long, sometimes slightly flattened at apex, smooth, usually glabrous, (3-)4-valved, pericarp 1-3 mm thick; seeds shaped like segment of an orange, ca. 1 cm long, completely surrounded by a fleshy sarcotesta, seed coat membranaceous.

Distribution: Mexico and C America and across northern S America, and also in the West Indies, Guyana and Suriname; in the Guianas a component of mixed lowland rain forest on non-flooded land; 6 collections studied (GU: 2; SU: 3).

Selected specimens: Guyana: Pomeroon R., Arunamai Cr., Forest Dept. 2527 (K). Suriname: Brownsberg, BW 6677 (FHO, NY, U).

Phenology: Flowers recorded in June and December.

Note: A widespread species containing several subspecies, only one of which *G. glabra* subsp. *glabra* occurs in Guyana and Suriname. The typical subspecies occupies most of the range of the species, but has not yet been recorded from French Guiana.

6. **Guarea gomma** Pulle, Recueil Trav. Bot. Néerl. 6: 271. 1909. Type: Suriname, Sectie O, BW 70 (holotype U). – Fig. 4

Tree, to 40 m tall, sometimes buttressed and with bole fluted up to 3 m above base; bark mid-brown, soft, slightly suberous and scaling in irregular patches. Young branches puberulous at first soon becoming glabrous, sometimes lenticellate. Leaves pinnate with or without an inactive terminal bud, all leaflets opening together; petiole 6-23 cm long, semiterete, strongly concave on upper surface at base and clasping young developing leaf buds, rachis 19-60 cm long, semiterete or terete, puberulous or glabrous; leaflets opposite or rarely subopposite, 8-26 pairs; petiolule 2-5(-10) mm long; leaflet blade chartaceous to subcoriaceous, narrowly oblong or oblanceolate (lower leaflets often elliptic), 4-30 × 3.3-7.4 cm, apex narrowly attenuate or short acuminate, base rounded to acute, cuneate, or attenuate, lower ones usually much smaller, glabrous or rarely with midrib puberulous abaxially, usually not glandular-punctate and -striate; venation eucamptodromous, midrib sunken adaxially, secondaries 12-18 pairs, ascending, parallel or slightly convergent, intersecondaries usually small and obscure, tertiaries oblique, parallel, moderately prominent abaxially. Inflorescence axillary, slender densely-flowered thyrse, 15-40 cm long, sometimes showing intermittent apical growth, lower branches 1-4 cm long, puberulous; pedicels 1-1.5 mm long. Flowers with calyx shallowly to deeply cyathiform, 2-3 mm long, with 3-5 irregular lobes to 3 mm long, sparsely appressed puberulous or glabrous; petals cream, 4-6, valvate or slightly imbricate, oblong to lanceolate, 8-12 × 2-3 mm, apex acute, densely appressed pubescent outside, glabrous inside; staminal tube 6.5-8 × 2-4 mm, margin undulate, glabrous, anthers 8-11, 1-1.5 mm long; nectary a stout stipe expanded to form an annulus below ovary, 1.5-2 mm long, glabrous; ovary densely appressed puberulous or pubescent, 4-6(-7)-locular, locules with 1-2 superposed ovules, style stout, puberulous or pubescent below. Capsule orange to red, globose to shortly pyriform, 2.5-3.2 × 2.5-2.8 cm, flattened at apex usually tapering to base, smooth, densely papillose-puberulous, 4-6-valved or sometimes less by abortion, valves 1-seeded, pericarp 2-3 mm thick; seeds slightly flattened dorso-ventrally, ca. 1.5-2 × 0.8-1.2 cm, surrounded by a thin, bright orange sarcotesta, seed coat thin but tough, hilum 1.2-1.6 × 0.7-1 cm.

Distribution: Ranging from the Guianas, S Venezuela to Colombia, Ecuador and Peru; in lowland non-flooded forest; 61 collections studied (GU: 4; SU: 24; FG: 16).

Selected specimens: Guyana: U. Takutu-U. Essequibo Reg., Rewa R., Clarke, 3625 (K); U. Essequibo R., Rewa R., near Corona Falls, Jansen-Jacobs et al. 5745 (K, U). Suriname: Coppename R. near Raleigh Falls, Pulle 324 (NY); Nickerie R., Paris Jacob Cr., Maas in LBB 11025 (NY); Jodensavanne-Mapane Cr. area (Suriname R.), Lindeman 6787 (U). French Guiana: Embouchure de la Crique Noire – Bassin du Sinnamary, Hoff 7380 (K); Saül, Mt. Galbao trail, between entrance to Grand Boeuf Mort trail and Cambrouse, Mori et al. 18671 (K, NY); Saül, Fumée Mt., Boom et al. 2099 (K, NY); Piste de St. Elie, Sabatier & Prévost 4111 (CAY).

Vernacular names: Suriname: gomma, doifisiri (Sranang), rode doifisiri, dyankoy maata, kodyo udu (Aucan: Saramaccan).

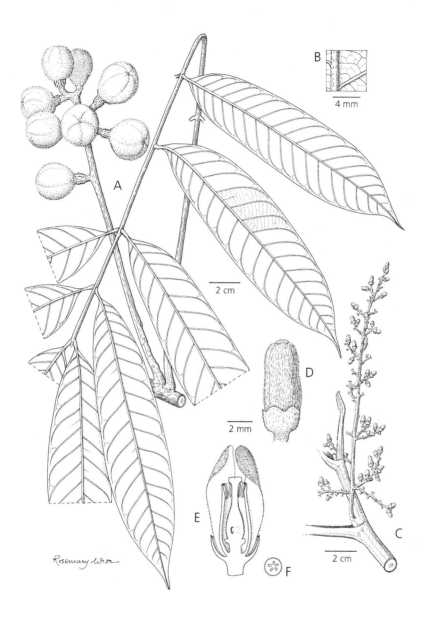

Fig. 4. *Guarea gomma* Pulle : A, habit with infructescence; B, detail of leaflet abaxial surface; C, inflorescence; D, flower bud; E, flower, longitudinal section; F, ovary, cross section. (A-B, Jansen-Jacobs *et al.* 5745; C-F, Clarke 3625).

Use: Leaves used in Suriname to cure rheumatism (van Andel 5404).

Phenology: Flowering from June to January.

Note: *Guarea gomma* is closely related to *G. macrophylla* subsp. *pachycarpa* and differs from it in a number of characters:

– *G. gomma*: leaflets ca. 4 times as long as broad; inflorescence up to 40 cm; petals 4-6, 8-12 mm long, anthers 1-1.5 mm long, ovary 4-6-locular; capsule 2.5-3.2 cm long, never tuberculate, valves 1-seeded.

– *G. macrophylla* subsp. *pachycarpa*: leaflets ca. 3 times as long as broad; inflorescence less than 25 cm; petals 4-5, 6-9 mm long, anthers usually less than 1 mm long, ovary nearly always 4-locular; capsule usually less than 2.5 cm long, sometimes tuberculate, valves with 1-2 superposed seeds.

7. **Guarea grandifolia** DC., Prodr. 1: 624. 1824. Type: Specimen annotated as "Cayenne ou Guyane françoise, Museum de Paris, 1821" (holotype G-DC). – Fig. 5

> *Guarea megantha* A. Juss., Bull. Sci. Nat. Géol. 23: 240. 1830; Mém. Mus Hist. Nat. 1: 241, 292. 1831. Type: French Guiana, Cayenne, Collector unknown s.n. (holotype P).

Tree, to 50 m tall, larger specimens usually with buttresses to 3-4 m tall; bark rough, pale brown or grey-brown, scaling in irregular plates, sometimes lenticellate. Young branches massive, indument varying from tomentose or velutinous to pubescent, or appressed puberulous, becoming glabrous. Leaves pinnate with a terminal bud showing intermittent growth; petiole 8-30 cm long, narrowly winged to semiterete, rachis 7-48(180) cm long, terete or channelled above, both tomentose to pubescent or puberulous, becoming glabrous; leaflets opposite, 10-22 pairs; petiolule 1-6(-15) mm long; leaflet blade chartaceous to coriaceous, usually oblong or narrowly elliptic, less frequently oblanceolate, 11-35 × (3.4-)4.5-10 cm, apex acuminate, acute, obtuse or truncate, base rounded, obtuse, acute, cuneate or attenuate, glabrous on adaxial surface, abaxial surface sparsely puberulous, sometimes intermixed with minute red papillae or glabrous, often glandular-punctate and -striate; venation eucamptodromous, midrib flat or slightly sunken adaxially, secondaries 13-22 pairs, shallowly ascending, straight or arcuate above, parallel or slightly convergent, the lower sometimes somewhat divergent, intersecondaries present or absent, tertiaries oblique, parallel. Inflorescence axillary or ramiflorous, 5-40 cm long, with indeterminate apical growth, rarely broadly pyramidal with lower branches to 15 cm long, puberulous to tomentose; pedicels 0.5-

2 cm long. Flower-buds oblong, obtuse; calyx reddish, cyathiform, 2-6 mm long, with 3-5 irregular acute, obtuse or rounded lobes of 2-3 mm long, appressed puberulous outside; petals white to cream, 4-5, valvate or slightly imbricate at apex, strap-shaped to lanceolate, 9-14 × 2-5 mm, acute, appressed pubescent to sericeous outside, glabrous inside; staminal tube 9-11 × 3-5 mm, margin truncate or undulate, glabrous or occasionally sparsely pubescent outside, anthers 10-12, 1.4-1.7 mm long; nectary a stout stipe expanded to form an annulus below ovary, 1-2.5 mm long, glabrous; ovary densely appressed pubescent to sericeous or strigose, 7-8-locular, locules with 2 superposed ovules, style appressed pubescent below, glabrous above, very stout and tapering gradually from base to apex. Capsule rich reddish brown, globose, ellipsoid, pyriform or obovoid, 5-6 × 3-4 cm, often tapering for ca. $^2/_3$ its length to a short stout stipe, sometimes apiculate or apex rounded or truncate, smooth or valves faintly longitudinally lined, densely puberulous or papillose, 5-8-valved, valves with 2 superposed seeds, pericarp 5-6 mm thick; seeds bright orange with a white hilum, ellipsoid when solitary, 1-2(-3) cm long, truncate at apex or base when superposed surrounded by a thin orange sarcotesta, hilum large, covering at least $^1/_3$ total seed surface.

Distribution: From Mexico through C America to northern S America, the Guianas and extending into central and western Amazonia; lowland non-flooded rain forest, often recorded from riverbanks; 26 collections studied (GU: 1; SU: 7; FG: 15).

Selected specimens: Guyana: U. Essequibo R., Rewa R, near Corona Falls, Jansen-Jacobs *et al.* 5782 (K, NY, U). Suriname: Coppename R. near Raleigh Falls, Voltzberg, Lanjouw 836 (NY); Nassau Mts., on slope along creek, Lanjouw *et al.* 2900 (K, NY); Nat. Res. Brownsberg, Brokopondo distr., Tjon Lim Sang in LBB 16224 (U). French Guiana: Mts. Bellevue de l'Inini, de Granville *et al.* 7628 (CAY, K, NY); Saül, between Eaux Claires and entrance to Grand Boeuf Mort, Mori *et al.* 22645 (NY); Orapu, km 7 route Fourgassié, Oldeman B-659 (NY).

Vernacular names: Suriname: grootbladige doifisiri, soort doifisiri. French Guiana: carapa (Cayenne).

Phenology: Flowering from July to December.

8. **Guarea guidonia** (L.) Sleumer, Taxon 5(8): 194. 1956. – *Samyda guidonia* L., Sp. Pl. ed. 1, 443. 1753. Type: Plate "Samyda foliis ovatis acuminates", Plumier in Burman, Pl. Amer. 6. 139, t. 147, f. 2. 1757. – Fig. 6

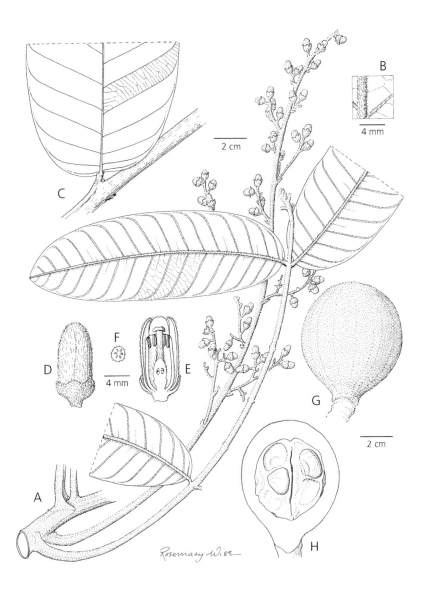

Fig. 5. *Guarea grandifolia* DC.: A, habit; B, detail of leaflet; C, leaflet base; D, flower bud; E, ½ flower; F. ovary section; G, fruit; H, fruit section. (A-B, Lanjouw 836; C-F Mori *et al*. 22645; G-H, Mori *et al*. 15685).

Guarea aubletii A. Juss., Bull. Sci. Nat. Géol. 23: 239. 1830. Type: French Guiana, Cayenne, Collector unknown (holotype P (herb. Richard), isotype P).

Tree, to 30 m tall, dioecious; bark fissured in older specimens. Young branches densely to sparsely puberulous or pubescent, soon becoming glabrous, typically smooth, dark brown with conspicuous pale lenticels. Leaves pinnate with a terminal bud showing intermittent growth, to 35(-45) cm long; petiole semiterete, rachis semiterete and canaliculate above or terete, puberulous or pubescent at first, soon glabrous; leaflets opposite, 9(-14) pairs; petiolule 1-6(-8) mm long; leaflet blade chartaceous, usually elliptic, oblong, or oblanceolate, rarely lanceolate, (10-)12.5-25(-35) × (2.5-)4-7(-10) cm, apex narrowly attenuate or acuminate less frequently acute, base usually acute, cuneate, or attenuate, glabrous, usually obscurely glandular-punctate and -striate; venation eucamptodromous, midrib slightly sunken, secondaries (6-)9-12(-17) pairs, arcuate ascending, parallel to slightly convergent, intersecondaries short, tertiaries oblique, parallel. Inflorescence usually axillary, usually a slender pyramidal thyrse, 10-25 cm long, with lower branches only 0.5-3 cm long, sparsely puberulous or pubescent; pedicels 1(-2) mm long or flowers sessile. Calyx patelliform or cyathiform, 1-2.5 mm long, with (2-)3-4(-5) acute to rounded or irregular lobes, sparsely appressed puberulous or pubescent outside; petals pale cream-coloured, 4-5, valvate or slightly imbricate, usually oblong, less frequently lanceolate, 5.5-7.5(-9.5) × 1-2 mm, apex acute, sparsely to densely appressed puberulous outside, papillose or glabrous inside; staminal tube cylindrical, 3.5-7 × 1.5-2.5(-3) mm, margin truncate or undulate, glabrous or occasionally slightly papillose, anthers (7-)8(-11), 0.75-1.25 mm long; antherodes narrower, not dehiscing, without pollen; nectary a stipe expanded above to form an annulus below ovary,1-2 mm long, glabrous; ovary sparsely to densely appressed pubescent, (3-)4(-5)-locular, locules 1-ovulate, style appressed puberulous or glabrous; pistillode similar, with well-developed abortive ovules. Capsule dark brown, globose to fig-shaped, 1.5-2.5 cm long, apex truncate, usually abruptly contracted at base into a short stout stipe, smooth not wrinkling on drying, shining, nearly always with conspicuous, sometimes pustular, pale lenticels, glabrous, 4-valved, valves 1-seeded, pericarp 1-2 mm thick, leathery; seeds ovoid, 1-1.3 × 0.6-0.8 cm, surrounded by a thin orange sarcotesta, seed coat thin, cartilaginous.

Distribution: Greater Antilles, Costa Rica and Panama and throughout tropical S America as far south as Paraguay and northern Argentina; characteristic species of riverbanks and periodically flooded forest; 134 collections studied (GU: 63; SU: 49; FG: 22).

Selected specimens: Guyana: Rupununi Distr., Kuyuwini Landing, Kuyuwini R., Jansen-Jacobs *et al.* 2373 (U). Suriname: Sipaliwini savanna area on Brazilian frontier, Oldenburger *et al.* 942 (U). French Guiana: Mt. Mahury base, E of Cayenne, Leeuwenberg 11614 (U); Ilet Maripa, route de l'Anse, Le Goff & Hoff 164 (CAY).

Vernacular names: Guyana: atïwa-u (C.), bastard crabwood, bat seed, buck-puke, buck vomit, crabwood, karababalli. Suriname: doifisirie, witbast-doifisiri, junkoimata, karababalli wedakoro abo (Ar.), karaballi (Ar.) kuleku (Car.). French Guiana: diankoimata.

Phenology: Flowering and fruiting throughout the year.

9. **Guarea kunthiana** A. Juss., Bull. Sci. Nat. Géol. 23: 240. 1830.
Type: French Guiana, Cayenne, Poiteau s.n. (holotype P-JU, isotypes K, G). – Fig. 7

Tree, to 30 m tall (often flowering as a small treelet of a few metres tall); bark greyish brown, scaling. Young branches often channelled, usually minutely appressed puberuluent, soon becoming glabrous, brown to greyish-brown, without lenticels. Leaves pinnate with a terminal bud showing some intermittent growth, 7.5-35 cm long; petiole semiterete, rachis channelled above, appressed puberulous or less frequently pubescent becoming glabrous; leaflets 2-6 pairs; petiolule 2-5 mm long; leaflet blade chartaceous to subcoriaceous, broadly to narrowly elliptic or oblanceolate, 15-25 × 5-10 cm, apex obtusely cuspidate to attenuate, obtuse, base usually acute, cuneate or attenuate, usually glabrous, less frequently sparsely to densely puberulous or pubescent abaxially, usually obscurely glandular-punctate and -striate; venation eucamptodromous, midrib flat, secondaries mostly 8-12 pairs, ascending, arcuate above, parallel or slightly convergent, intersecondaries usually absent, tertiaries widely spaced and obscure. Inflorescence usually axillary, indeterminate with intermittent growth from apex, a slender to broadly pyramidal thyrse, 3-5-25 cm long, with lower branches to 12 cm long, minutely puberulous to pubescent; pedicels 2-3 mm long. Flowers: calyx cyathiform to patelliform, 1.5-3 mm long, with (3-)4 irregularly rounded, obtuse acute lobes, 0.5-2.5 mm long or margin truncate, sparsely appressed puberulous outside; petals cream-coloured, (3-)4, valvate, oblong to lanceolate, 7-12 × 1.5-4 mm, apex acute and often hooded, densely appressed puberulous or appressed pubescent outside, glabrous inside; staminal tube 5.5-10 × 2-3 mm, margin truncate, undulate, or crenulate, glabrous or sparsely to moderately appressed puberulous outside, glabrous inside, anthers (7-)8(-10), 1.2-1.8 mm long; nectary broadly stipitate, sometimes expanded to

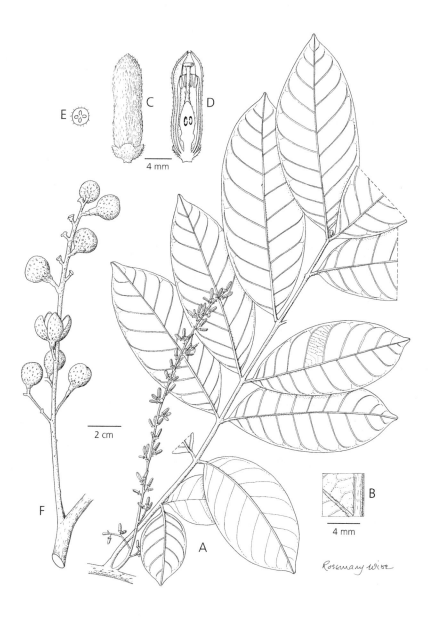

Fig. 6. *Guarea guidonia* (L.) Sleum.: A, habit; B, detail of leaflet abaxial surface; C, flower bud; D, flower, longitudinal section; E, ovary, cross section; F, infructescence. (A-E, Hahn & Tuvari 5213; F, Hoffman *et al.* 1526).

form an annulus below ovary, 0.5-2.5 mm long, glabrous; ovary ovoid, glabrous, (3-)4-locular, locules with 2 superposed ovules, style glabrous. Capsule reddish, ellipsoid to globose or irregularly shaped, 2.5-4.5 × 1.5-3.5 cm, apex rounded to acute, sometimes apiculate, base acute to rounded sometimes with a very short stout stipe, smooth or rarely obscurely verrucose, usually with large pale lenticels, glabrous, 4-valved, valves with 2 superposed seeds, pericarp 1-4 mm thick, leathery; seeds 1-2 in each valve, to 2.5 × 1.8 cm, surrounded by a fleshy orange sarcotesta, seed coat thick, woody.

Distribution: From Costa Rica and Panama throughout tropical S America, including the Atlantic coastal forests of Brazil to Paraguay and Bolivia; a species of lowland rain forest usually on non-flooded land; 55 collections studied (GU: 6; SU: 28; FG: 24).

Selected specimens: Guyana: U. Takutu-U. Essequibo Reg., Acarai Mts., 4 km S of Sipu R., Clarke *et al.* 7677 (U). Suriname: Nickerie R., Paris Jacob Cr., Maas in LBB 11024 (K). French Guiana: Saül, Mts. la Fumée, Boom & Mori 2105 (K); Saül, route de Belizon, Mori & Pipoly 15615 (CAY).

Vernacular names: Suriname: doifisiri (Sranang), redi doifisiri, zwarte doifisiri or zwarte doifiesierie, sali, warakabadan djamaro.

Phenology: Flowering and fruiting throughout the year.

10. **Guarea macrophylla** Vahl, Eclog. Am. 3: 8. 1807. Type: Lesser Antilles, St. Croix, Montserrat, Ryan s.n. (holotype C).

Tree, to 20 m tall. Young branches puberulous or pubescent, becoming glabrous, pale to mid-brown, usually without lenticels. Leaves pinnate with a dormant terminal bud, 13-20 cm long; petiole semiterete, rachis semiterete or channelled above, indument sparsely pubescent, becoming glabrous; leaflets opposite, 3-4 pairs; petiolule 3-6 mm long; leaflet blade chartaceous to coriaceous, elliptic to oblong, 10.5-13 × 3.5-4.5 cm, lower leaflets often much smaller, apex attenuate or acuminate, base acute or narrowly attenuate, glabrous adaxially, sparsely pubescent to glabrous abaxially, sometimes with granular red papillae, sometimes glandular-punctate and -striate; venation eucamptodromous, midrib slightly sunken adaxially, secondaries 7-10 pairs, moderately to steeply ascending, slightly arcuate, slightly convergent, intersecondaries short or absent, tertiaries oblique. Inflorescence axillary, a slender thyrse, 5-25 cm long, with lower branches 0.5-1.5 cm long, sparsely puberulous; pedicel 0.5-1.5 mm long. Flowers: calyx cyathiform, 1-3 mm long, with 3-4 acute, obtuse or

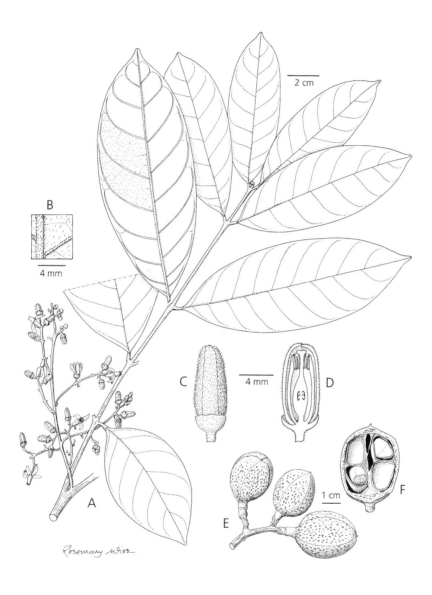

Fig. 7. *Guarea kunthiana* A. Juss.: A, habit; B, detail of leaflet abaxial surface; C, flower bud; D, flower, longitudinal section; E, fruit; F, fruit section. (A-D, Mori & Pipoly 15615; E-F, Mori & Gracie 18330).

rounded lobes, sparsely appressed puberulent on outer surface; petals 4-5, valvate, strap-shaped, 6-9 × 1-2.5 mm, apex acute, appressed puberulous on outer surface, glabrous inside; staminal tube 4.5-7 × 1.5-3 mm, margin entire or undulate, glabrous, anthers 8, 0.7-1.2 mm long; nectary a stout stipe expanded to form an annulus below ovary, 0.5-2 mm long, glabrous; ovary tapering to apex, appressed puberulous, 4(-5)-locular, locules with 1-2 superposed ovules, style stout, glabrous in upper half. Capsule globose or pyriform, 1.5-2.5 cm, apex emarginate, truncate or rounded, often tapering from near apex to base and finally contracted into a short stipe, valves obscurely tuberculate or smooth, sometimes becoming prominently wrinkled on drying, minutely puberulous, 4-valved, valves with 1-2 superposed seeds, pericarp 1-2 mm thick; seeds 0.6-1.5(-2) cm long, surrounded by a thin orange sarcotesta, seed coat thin, cartilaginous.

Note: Species with 5 subspecies, 2 of these subspecies occur in the Guianas.

KEY TO THE SUBSPECIES

1　Inflorescence 5-10 cm long; fruit fig-shaped, valves obscurely tuberculate.
. . . . .10a. *G. macrophylla* subsp. *pachycarpa*
　Inflorescence 12-25 cm long; fruit fig-shaped to globose, valves smooth . .
. . . . .10b. *G. macrophylla* subsp. *pendulispica*

10a. **Guarea macrophylla** Vahl subsp. **pachycarpa** (C. DC.) T.D. Penn., Fl. Neotrop. 28: 289. 1981. – *Guarea trichilioides* L. var. *pachycarpa* C. DC. in Mart., Fl. Bras. 11(1): 184. 1878. Type: Brazil, Pará or Maranhão, Martius s.n. (holotype M).

Flowers scented; calyx red; corolla varying from white to coral-pink. Capsule deep red-purple.

Distribution: Venezuela, French Guiana, Amazonian and coastal Brazil, Amazonian Peru and Bolivia; in riverside vegetation in lowland rain forest; 7 collections studied (SU: 2; FG: 5).

Selected specimens: Suriname: Brokopondo, Brownsberg, Troon LBB 15224 (U). French Guiana: Inselbergs de la Haute Wanapi, Engel & Tarcy 4 (CAY); Rivière Compte, 75 km S of Cayenne, Oldeman 1221 (K, U).

Vernacular name: Suriname: doifisiri.

Phenology: Fruiting in March.

10b. **Guarea macrophylla** Vahl subsp. **pendulispica** (C. DC.) T.D. Penn., Fl. Neotrop. 28: 290. 1981. – *Guarea pendulispica* C. DC. in Herzog, Feddes Repert. 7: 59. 1909. Type: Bolivia, Cunucu, Herzog 313 (holotype G).

Flowers cream or rose coloured. Capsule dark red to purplish.

Distribution: Costa Rica, Panama to Venezuela and Guyana, Colombia to Bolivia; a tree of non-flooded rain forest; 1 collection studied (GU: 1).

Specimen examined: Guyana: Rupununi Distr., Kuyuwini Landing, Jansen-Jacobs *et al.* 2487 (K, U).

Phenology: Fruiting in February.

11. **Guarea michel-moddei** T.D. Penn. & S.A. Mori, Brittonia 45(3): 231. 1993. Type: French Guiana, Saül, La Fumée Mt. Trail, Mori *et al.* 20963 (holotype NY, isotypes CAY, K).

Tree, to 6 m tall, unbranched; bark thick, fissured, suberous. Young shoots golden tomentose or villous at first, soon becoming glabrous and suberous. Leaves pinnate, in terminal, spirally arranged clusters, 30-70 cm long, with a terminal bud with intermittent growth; petiole 8-16 cm long, petiole and rachis channelled above, sparsely hirsute; leaflets 3-10 pairs; petiolule 3-7 mm long; leaflet blade oblong, 22-41 × 5-11 cm, basal leaflets smaller, apex obtuse to narrowly attenuate, base acute to obtuse, glabrous adaxially, midrib and veins sparsely hirsute abaxially; venation eucamptodromus or brochidodromus in upper half, venation strongly impressed adaxially and prominent abaxially (bullate), secondaries 15-18 pairs, straight or slightly arcuate, parallel or slightly convergent, ascending, intersecondaries short or absent, tertiaries oblique. Inflorescence a few-flowered, congested thyrse to 1 cm long, arising from trunk; pedicels to 1 mm long, stout. Flowers: calyx pink, cyathiform, 2-3 mm long, irregularly 4-lobed, lobes ca. 1.5 mm long, sparsely puberulous outside; petals cream, 4, valvate, broadly oblong, ca. 13 × 3.5 mm, acute, reflexed at apex, shortly appressed pubescent outside; staminal tube ca. 9.5 × 4 mm, glabrous, margin undulate, anthers 8, ca.1.5 mm long; nectary with a broad stipe, ca. 1 mm long, expanded to form annular ring below ovary; ovary densely strigose, 6-8-locular, locules with 2 superposed ovules, style glabrous, slightly exceeding staminal tube, stigma broad, discoid. Capsule deep red, globose to depressed globose, 3-4 cm long, with anastomosing and convoluted fleshy ribs or wings of ca. 1 cm deep; pericarp (not including wings) ca. 2 mm thick; seeds 2, superposed in each valve, 1.3-1.5 × 1 cm, surrounded by fleshy orange sarcotesta.

Distribution: French Guiana and adjacent Brazil; in lowland rain forest where it appears to be confined to ridge tops and slopes, up to 600 m alt; 10 collections studied (FG: 10).

Selected specimens: French Guiana: Mt. Bellevue de l'Inini, de Granville *et al.* 7749 (K); Saül, La Fumée Mt., Pennington *et al.* 12110 (K); Mt. de Kaw, Domaine Tresor, Ek *et al.* 1312 (U); Nouragues Reserve, Delnatte *et al.* 1479 (CAY).

Phenology: A single flowering specimen was gathered in September, fruit has been collected from January to June.

Note: It is not yet known if the flowers of this species are bisexual or unisexual. The few flowers available for study do not show any obvious differences in structure. Most species of *Guarea* however, have unisexual flowers with rudiments of the opposite sex. Although the fruit of *G. michel-moddei* dehisces, the valves do not open widely and the seeds are retained within the fruit. In other species of *Guarea*, the usually dull-coloured fruits open widely and the seeds, covered by an orange sarcotesta, are suspended below the fruit on a long funicle. In *G. michel-moddei*, the visual attraction for dispersers is provided by the bright red fleshy wings of the fruit. Seed dispersers may be primates as in the morphologically similar *Guarea cristata* T.D. Penn., which is dispersed by the capuchin monkey (*Cebus paella*).

12. **Guarea pubescens** (Rich.) A. Juss., Bull. Sci. Nat. Géol. 23: 240. 1830. – *Trichilia pubescens* Rich., Actes Soc. Hist. Nat. Paris 1: 108. 1792. Type: French Guiana, Cayenne, Leblond 61 (holotype G, isotype fragm. P).

Tree, 3-5(10) m tall, dioecious or hermaphroditic, in early stages of growth monopodial and usually unbranched; bark pale grey or greyish white, strongly suberized, fissured (subsp. *pubescens*), (subsp. *pubiflora* produces aerial adventitious roots when subject to flooding). Young branches sparsely to densely coarse pubescent, becoming glabrous, hairs often yellowish, pale grey-white with suberous bark. Leaves pinnate with a terminal bud showing intermittent growth sometimes, to 30 cm long; petiole and rachis semiterete or narrowly winged, wings broader below insertion of leaflets, often channelled above, coarsely pubescent at first, becoming glabrous; leaflets opposite, 2-7 pairs; petiolule 1-2 mm long; leaflet blade chartaceous, elliptic, oblong, or oblanceolate, 9-20 × 4-7 cm, apex usually attenuate or acuminate, base acute, cuneate, or attenuate, usually glabrous except for a few hairs on abaxial surface sometimes glandular-punctate or -striate, adaxial surface often with minute raised dots; venation eucamptodromous, secondaries 8-12 pairs, usually arcuate and convergent, intersecondaries usually short or absent, tertiaries oblique and parallel, often numerous (to 10 per cm) and prominent abaxially. Inflorescence axillary or ramiflorous, a few-flowered slender raceme or thyrse, 2-15 cm long, with a few short branches near base, sometimes terminated by a cluster of small bracteoles and showing intermittent growth, coarsely pubescent to subglabrous; pedicels 1-2.5 mm long. Flowers unisexual or bisexual; calyx often closed in bud and splitting irregularly to form 3-5 usually ovate acute lobes, mature calyx patelliform or cyathiform, 2-3.5 mm long, often lobed to base, sparsely and coarsely appressed pubescent outside or subglabrous; petals 4(-5), valvate, oblong to lanceolate, 7-9 × 2-3 mm, apex acute, appressed pubescent to strigose outside, glabrous inside; staminal tube 5-7(-9) × 1.5-3 mm, margin undulate, sparsely appressed pubescent or glabrous on both sides, anthers (7-)8(-9), 0.7-1.3 mm long; antherodes similar but narrower, not dehiscing, without pollen; nectary a stipe expanded to form an annulus below ovary, 0.5-1.5 mm long, glabrous; ovary rounded, shortly ovoid, densely long-strigose, 4(-5)-locular, locules 1-ovulate, style glabrous. Capsule dark red or purple, depressed-globose to globose or pyriform, 1.2-2 cm diam., usually broader than long, 4-valved, tapering at base to a short stipe or truncate, valves obscurely to prominently 3(-4)-ribbed, often minutely tuberculate between ribs or less frequently smooth, valves 1-seeded, pericarp 0.5-1.5(-2) mm thick, puberulous intermixed with scattered coarse long hairs, leathery-fleshy; seed shaped like segment of an orange, 0.8-1.5 × 0.6-1.0 cm, surrounded by a thin, orange sarcotesta, testa thin but tough, hilum 0.5-1 × 0.25-0.5 cm.

KEY TO THE SUBSPECIES

1 Petiole and rachis unwinged; midrib usually sunken and slightly puberulous, upper surface with minute raised dots; capsule usually globose to fig-shaped, minutely tuberculate, ribbed or not . . . .
. . . . .12a. *G. pubescens* subsp. *pubescens*
Petiole and rachis narrowly winged; midrib prominent on upper surface, glabrous, upper surface without raised dots; capsule depressed globose, irregularly ribbed . . . . 12b. *G. pubescens* subsp. *pubiflora*

12a. **Guarea pubescens** (Rich.) A. Juss. subsp. **pubescens** – Fig. 8 A-F

Guarea affinis A. Juss., Bull. Sci. Nat. Géol. 23: 240. 1830. Type: French Guiana, Perrottet s.n. (holotype P).
Guarea richardiana A. Juss., Bull. Sci. Nat. Géol. 23: 240. 1830. Type: "Guiana, between Conana and Approngas", Collector unknown s.n. (holotype P).

Guarea davisii Sandwith, Kew Bull. 330. 1933. Type: Guyana, Apoteri, Rupununi R., Davis in Forest Dept. 2092 (holotype K, isotype FHO).
Guarea concinna Sandwith, Kew Bull. 309. 1948.Type: Guyana,Tumatumari, Gleason 443 (holotype K, isotypes GH, NY).

Distribution: Northern S America from western Venezuela through the Guianas to Brazil (Amapa) and also in the western and central Amazon basin; a tree of non-flooded lowland forest; 108 collections studied (GU: 24; SU: 22; FG: 62).

Selected specimens: Guyana: Potaro-Siparuni Reg., Iwokrama Rain Forest Reserve, Clarke & Hoffman 606 (U); Rupununi Distr., foothills of NW Kanuku Mts., near Moco-Moco, Maas *et al.* 3800 (U). Suriname: Mapane Cr. area, Elburg in LBB 13506 (U); Jodensavanne-Mapane Cr. area, Schulz in LBB 8590 (U). French Guiana: Kaw Mts., Degrad Fourgassié, Cremers 11113 (CAY); Mt. Bellevue de l'Inini, de Granville 7876 (U); Papaichton, Lawa R., Sastre *et al.* 8152 (U).

Vernacular names: Guyana: kufiballi (Arawak). Suriname: banjabo, doifisiri, rode doifisiri. French Guiana: diankoimata (Par.), kaliakulaka a (Wayapi), kalaman.

Phenology: Flowering and fruiting throughout the year.

Note: Specimens recorded from riverbanks in French Guiana are somewhat intermediate between subsp. *pubescens* and subsp. *pubiflora*.

12b. **Guarea pubescens** (Rich.) A. Juss. subsp. **pubiflora** (A. Juss.) T.D. Penn., Fl. Neotrop. 28: 294. 1981. – *Guarea pubiflora* A. Juss., Bull. Sci. Nat. Géol. 23: 240. 1830. Type: Brazil, Amazonas, Rio Negro, Barcellos, Collector unknown s.n. (holotype P).   – Fig. 8 G-H

Distribution: Southern Venezuela, French Guiana and central and southern Amazonian Brazil; confined to riverbanks and periodically or permanently flooded forest; 7 collections studied (FG: 6).

Selected specimens: French Guiana: Le Tampoc, above Saut Koumaiou Soula, Cremers 4505 (CAY, K); Bas Camopi, de Granville 1330 (K); R. Yaroupi, 8 km above Saut Tainous, Oldeman 530 (K).

Phenology: Flowering April-May.

13. **Guarea scabra** A. Juss., Bull. Sci. Nat. Géol. 23: 239. 1830. Type: French Guiana, Cayenne, Collector unknown s.n. (holotype P-JU, isotype P).   – Fig. 9

Tree, to 15(25) m tall, dioecious, trunk sometimes with prominent bosses; bark mid-brown, scaling in slightly suberous pieces in older specimens, bark of younger branches rather rough and sometimes slightly suberous. Young branches minutely appressed puberulous, soon becoming glabrous. Leaves pinnate with a terminal bud showing intermittent growth; petiole 4.5-12 cm long, semiterete, rachis 9-40(80) cm long, terete or canaliculate above, both appressed puberulous at first soon glabrous; leaflets opposite, 6-12 pairs; petiolule 2-4(-5) mm; leaflet blade chartaceous, elliptic, oblong, or oblanceolate, 12.5-22.5 × 4-7.5 cm, lower leaflets often shorter and broader, apex attenuate to long acuminate less frequently obtusely acuminate, base acute, cuneate or attenuate, glabrous, but often with minute red papillae on abaxial surface, not glandular-punctate or -striate; venation eucamptodromous, midrib flat or slightly sunken on adaxial surface, secondaries (7-)9-12(-13) pairs, ascending, arcuate, convergent, intersecondaries absent, tertiaries oblique, obscure. Inflorescence axillary and on smaller branches, broadly pyramidal, 3.5-12 cm long, lower branches to 10 cm long, weak slender axes, spreading and laxly-flowered, subglabrous; pedicels (1.5-)2 mm long. Calyx reddish, patelliform, 1-1.5 mm long, margin truncate or with 4 irregular lobes to 0.75 mm long, sparsely and minutely appressed puberulous to subglabrous; petals cream-coloured in lower half and tipped with pink, 4-5, valvate or slightly imbricate, oblong or lanceolate, (7-)9.5-12 × 2-3 mm, apex acute, minutely appressed puberulous outside, glabrous inside; staminal tube 7.5-9.5 × 2-3 mm, margin undulate or truncate, glabrous, anthers 8-9, 1-1.25 mm long; antherodes smaller, ca. 0.75 mm long, not dehiscing, without pollen; nectary a stipe expanded to form an annulus below ovary, 1-2 mm long, glabrous; ovary strigose, 4(-5)-locular, locules mostly 1-ovulate, style glabrous above; pistillode similar, with well-developed non-functional ovules. Capsule globose to broadly pyriform, 2-2.2 cm long, slightly constricted between seeds on drying, apex truncate, base rounded or tapered, 4-valved, valves 1-seeded, smooth, minutely puberulous mixed with some longer pubescence, pericarp 2-3 mm thick; seeds ca. 1 cm long, surrounded by a thin orange sarcotesta, seed coat thin and hard.

Distribution: The Guianas, and adjacent Amapá, Brazil, to central and western Amazon basin; a tree of non-flooded lowland forest; 20 collections studied (GU: 5; SU: 7; FG: 7).

Selected specimens: Guyana: Potaro-Siparuni Reg., Kaieteur National Park, along Potaro R., upstream of Kaieteur F., Gillespie et al. 1377 (K, US); U. Takutu-U. Essequibo Reg., Kwitaro R., Clarke 6311 (K, US); Gunn's, Essequibo R., Jansen-Jacobs et al. 1912 (K, U). Suriname: Vicinity of Ulemari R., ca. 150 km upstream from its confluence with

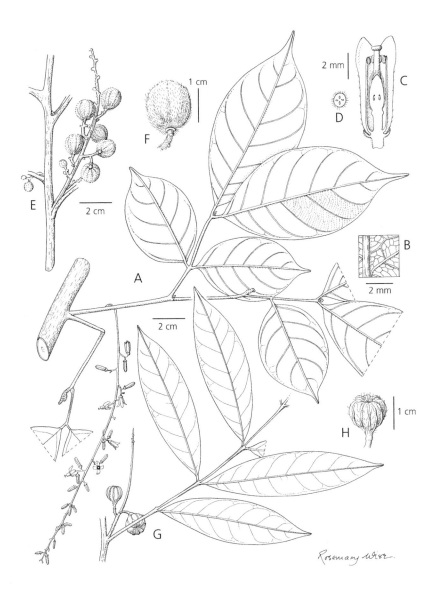

Fig. 8. *Guarea pubescens* (Rich.) A. Juss.ssp. pubescens : A, habit; B, detail of leaflet abaxial surface; C, flower, longitudinal section; D, ovary, cross section; E, infructescence; F, fruit; ssp pubiflora (A.Juss.) T.D. Penn.: G, habit; H, fruit. (A-D, Mori & Pipoly 15626; E-F, Jenman 316; G-H, Acevedo-Rodriquez 7967).

Litani R., Hammel *et al.* 21662 (K, MO); Boven-Tapanahony R., Paloemeu R., Schulz in LBB 8151 (U); Tumuc-Humac Mts., Suriname-Brazil border, Jari R., Acevedo-Rodriguez 6090 (K, US). French Guiana: Matiti, Prévost *et al.* 4390 (CAY, K, P); Mt. Bellevue de l'Inini, de Granville 7566 (CAY, K); Saül, Layon Biche, Route de Belizon, ca. 1 km from Eaux Claires, Mori *et al.* 21165 (K, NY); Inselbergs of haute Wanapi, de Granville *et al.* 15977 (CAY).

Vernacular names: Suriname: doifisirie (N.E.). French Guiana: kamwi (Palikur), kawapkamwi (Palikur).

Phenology: Flowering from August to October and fruiting from November to June.

Note: In its vegetative features this species is similar to *G. convergens* and can be easily confused in the absence of flowers or fruit. However, the inflorescence and fruit enable them to be separated without much difficulty. The chief distinguishing features are:

G. scabra: inflorescence branched, broadly pyramidal, and lax-flowered; petals (7-)9.5-12 mm long; capsule globose to pyriform to 2 cm diam., valves with 3-4 prominent narrow ribs.

G. convergens: inflorescence narrow, densely-flowered, with few short branches; petals 6.5-7.5 mm; capsule globose, 3.5-4 cm diam., valves obscurely 6-7-ribbed.

14. **Guarea silvatica** C. DC. in Mart., Fl. Bras. 11(1): 195. 1878. Type: Brazil, Amazonas, in sylvis Japurensibus, Martius s.n. (holotype M).

Tree, to 20 m tall (often flowering as a small treelet), dioecious (?). Young branches glabrous (young leaf buds with sparse minute puberulous indument), greyish-green, smooth, becoming pale brown, sometimes with longitudinal lenticels and often scaling in thin irregular pieces to reveal a green undersurface. Leaves pari-pinnate without a terminal bud, 12-20 cm long; petiole and rachis terete, glabrous; leaflets 2-4 pairs; petiolule 3-4 mm long; leaflet blade chartaceous to subcoriaceous, usually elliptic, 10-21 × 4-8 cm, apex acuminate, base narrowly attenuate or acute, glabrous, not glandular-punctate or -striate; venation usually eucamptodromous less frequently brochidodromous, midrib slightly prominent, secondaries 6-10 pairs, arcuate ascending, usually convergent, intersecondaries often long, tertiaries forming a prominent reticulum. Inflorescence axillary, 8-40 cm long, nearly always unbranched, flowers clustered along axis in distant cymose fascicles, glabrous; pedicels 2-4 mm long. Flowers: calyx rotate or patelliform, 0.5-1 mm long, margin entire or with 4 short lobes,

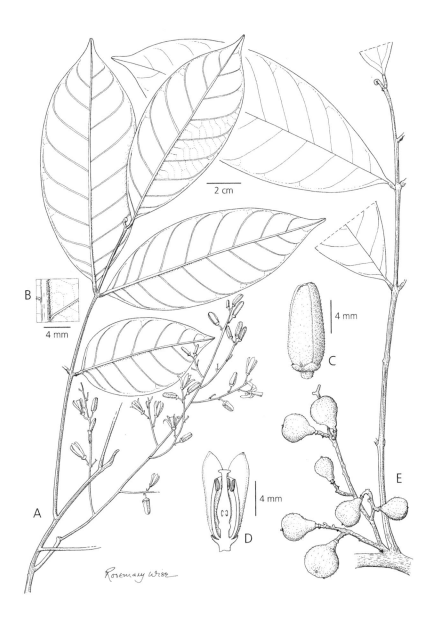

Fig. 9. *Guarea scabra* A. Juss.: A, habit; B, detail of leaflet abaxial surface; C, flower bud; D, flower, longitudinal section; E, infructescence. (A-B, Acevedo-Rodriquez *et al.* 5753; C-D, Mori *et al.* 24090; E, Evans & Peckham 2859).

glabrous; petals cream-coloured, 4(-5), imbricate, strap-shaped, 5-7.5 × 1-3 mm, apex rounded to acute, usually glabrous; staminal tube (4-)5-6 × 1.5-3 mm, margin entire, undulate or shallowly lobed, glabrous, anthers 8(-10), 0.5-0.7 mm long; nectary with stout stipe expanded to form a collar beneath ovary, 1-2 mm long, glabrous; ovary ovoid, glabrous, 2-3-locular, locules with 1-2 superposed ovules, style stout, glabrous. Capsule constricted between seeds, smooth, glabrous, 3.8-5.0 × 2.5-4 cm, 2-valved, valves 1-seeded, pericarp 1.5-2 mm thick; seeds ovoid to ellipsoid, 2.7-3 × 1.8-2 cm, surrounded by a thin fleshy orange sarcotesta, hilum large, ca. 1.8-2 × 0.9 cm, seed coat bony, ca. 1 mm thick.

Distribution: Southern Venezuela and the Guianas, Amazonian Colombia, Ecuador, Peru and Amazonian Brazil; in lowland rain forest on non-flooded land, and also persisting in cut-over forest; 18 collections studied (GU: 9; SU: 1; FG: 7).

Selected specimens: Guyana: U. Takutu-U. Essequibo Reg., Wassarai Mts., 12 km S of S. Kassikaityu R., Clarke *et al.* 8681 (K), 4597 (U). Suriname: Tumuc Humac Mts., Talouakem, Acevedo-Rodriguez *et al.* 5986 (K). French Guiana: Site ONF de Mt. Tortue, Sabatier *et al.* 4714 (K); 6 km NNW of Saül, vicinity of Les Eaux Claires, N.P. Smith *et al.* 13 (K); Massif des Emerillons, Molino & Sabatier 2491 (CAY).

Phenology: Flowering from September to January and fruiting from May to September.

15. **Guarea trunciflora** C. DC. in A. DC. & C. DC., Monogr. Phan. 1: 571. 1878. Type: Plate in Poeppig & Endlicher, Nov. Gen. Sp. Pl. 3: 39, t. 245. 1843. – Fig. 10

Tree, to 20 m tall, dioecious; bark soft brown, rather suberous, scaling or slightly fissured. Young branches stout, golden tomentose at first becoming suberous, fissured longitudinally. Leaves pinnate with a terminal bud showing intermittent growth, to 35 cm long; petiole semiterete, rachis terete or channelled above, tomentose, becoming lenticellate when older; leaflets to 6 pairs; petiolule 2-4 mm long; leaflet blade subcoriaceous, elliptic to oblanceolate, 15.5-18 × 5-7 cm, lower often much shorter and broader, apex shortly acuminate, or obtusely cuspidate, base acute, adaxial surface glabrous except for pubescent midrib, abaxial surface uniformly puberulous with crisped hairs intermixed with minute red papillae, not glandular-punctate or -striate; venation eucamptodromous, midrib sunken, secondaries 13-16 pairs, ascending, slightly arcuate, parallel, intersecondaries absent, tertiaries obscure, oblique, parallel. Inflorescence axillary or in axils of fallen leaves, 3-20 cm long, a laxly-branched pyramidal thyrse with lower branches widely spreading to 6 cm long; flowers subtended by 1-3 lanceolate bracteoles of 4-6 mm long, densely golden pubescent or tomentose; pedicels 1-2 mm long. Calyx deeply cyathiform to short cylindrical, 7-8 mm long, closed in bud and splitting irregularly to form 2-4 acute to rounded lobes to 6 mm long, densely crisped pubescent outside; petals 4(-5), valvate, oblong or strap-shaped, 11-14 × 2-5 mm, apex acute, densely golden sericeous outside, glabrous within; staminal tube 8.5-10 × 3-3.5 mm, margin undulate, glabrous, anthers 8-10, 1.1-1.5 mm long; antherodes similar, narrower, not dehiscing, without pollen; nectary a stipe expanded to form an annulus below ovary, 1-2 mm long, glabrous; ovary densely strigose, 4-5-locular, locules with 2 superposed ovules, style glabrous above; pistillode similar, with well-developed but abortive ovules. Capsule pyriform, 3.5-5 × 3-3.5 cm, 4-5-valved, smooth or with faint longitudinal lines when dry, shortly velutinous, valves with 2 superposed seeds, pericarp 5-7 mm thick; seeds ca. 1.7 × 1 cm, with plano-convex, superposed cotyledons.

Distribution: Amazonian Colombia and Peru, southern Venezuela and Guyana, French Guiana to central Amazonian Brazil; in non-flooded rain forest on brown sand; 7 collections studied (GU: 4; FG: 1).

Selected specimens: Guyana: U. Takutu-U. Essequibo Reg., Acarai Mts, 4 km S of Sipu R., Clarke *et al.* 7117 (K). French Guiana: Savanne Roche Virginie, Sabatier & Gonzalez 5399 (CAY).

Phenology: Fruiting in August.

16. **Guarea velutina** A. Juss., Bull. Sci. Nat. Géol. 23: 240. 1830. Type: Brazil, Collector unknown, annotated "Herbar Lusitanicum, Brésil" (holotype P).

Tree, to 30 m tall. Young branches velutinous or tomentose, indument persistent, with scattered black, extrafloral nectaries of 0.4-0.6 mm long. Leaves pinnate with a terminal bud showing intermittent growth; petiole 8-11 cm long, semiterete, rachis 3-47 cm long, terete, velutinous becoming subglabrous; leaflets 2-11 pairs; petiolule 2-4 mm long; leaflet blade coriaceous, elliptic, oblong or lanceolate, 6.3-21 × 2.5-7.3 cm, apex attenuate or short acuminate, base (at least upper ones) usually truncate or obtuse often slightly asymmetric, adaxial surface glabrous or sparsely puberulous or midrib puberulous, abaxial surface uniformly and often densely puberulous with crisped hairs intermixed with minute red papillae, often drying dark brown below, sometimes obscurely glandular-punctate and -striate; venation eucamptodromous, midrib on

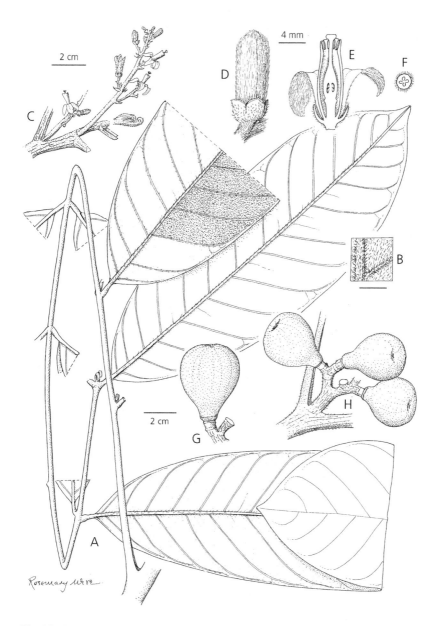

Fig. 10. *Guarea trunciflora* C.DC.: A, habit; B, detail of leaflet abaxial surface; C, inflorescence; D, flower bud; E, flower, longitudinal section; F, ovary, cross section; G, mature fruit. (A-B, G, Clarke *et al*. 7719; C-E, Mori *et al*. 21325; F, Clarke 3627).

adaxial surface flat, secondaries 7-12 pairs, ascending, straight, parallel or slightly convergent, intersecondaries sometimes present, to half length of secondaries, tertiaries oblique or obscure. Inflorescence axillary, a pyramidal thyrse, 4-16 cm long, lower branches to 8.5 cm long, densely pubescent; pedicels 0.5-1.5 mm long. Flowers greenish white to cream coloured; calyx patelliform, ca. 2 mm long with 3-4 acute, obtuse, or rounded lobes of 0.5-1 mm long, sparsely appressed puberulous outside; petals 3-5, valvate or slightly imbricate at apex, oblong, 7-9 × 1.5-2.5 mm, apex acute, appressed pubescent outside, glabrous within; staminal tube 6-7.5 × 2-2.5 mm, margin undulate or truncate, pubescent or glabrous outside, glabrous inside, anthers 8-9, 1-1.3 mm long; nectary with a stipe expanded to form an annulus below ovary, 1-1.5 mm long, glabrous; ovary prominently ribbed, densely pubescent or strigose, 4-5-locular, locules 1-ovulate, style glabrous. Capsule green with red wings, globose, 2.5-3 cm, apex truncate, base rounded, bearing 12-16 thin convoluted and anastomosing longitudinal wings, 5-7 mm deep, puberulous, hairs intermixed with numerous red papillae, 4-valved, valves 1-seeded, pericarp (excluding wings) 3-5 mm thick; seeds with orange aril, 1.5 cm long.

Distribution: Amazonian Colombia and Brazil, Guyana; non-flooded lowland forest, up to 750 m alt.; 4 collections studied (GU: 4).

Selected specimens: Guyana: U. Takutu-U. Essequibo Reg., Kuyuwini R. trail to Kassikaityu R., Clarke 4461 (K); S Kassikaityu R., transect S. Kassikaityu R. to Wassarai Mts., Clarke et al. 7932 (K); Wassarai Mts., summit of unnamed peak, Clarke et al. 8056 (K).

Phenology: Specimens with flower buds were collected in August and fruiting specimens in May to August.

6. **KHAYA** A. Juss., Bull. Sci. Nat. Géol. 23: 238. 1830.

Type: *K. senegalensis* (Desr.) A. Juss. (Swietenia senegalensis Desr.)

Deciduous trees, monoecious. Shoot apex bearing a cluster of bud-scales. Leaves paripinnate, glabrous. Flowers unisexual. Inflorescences axillary, much-branched thyrses; calyx 4-5-lobed, lobes rounded, imbricate; petals 4-5, free, much longer than calyx in bud, contorted; staminal tube urceolate or cup-shaped, terminated by 8-10 small lobes, anthers 8-10, included in throat of tube, alternating with lobes; nectary in male flowers cushion-shaped, in female flowers reduced to an indistinct swelling at base of ovary; ovary 4-5-locular, locules with 12-16 ovules, style-head thick, discoid; pistillode with vestigial ovules. Fruits erect, subglobose,

woody, septicidal capsules, opening by 4-5 valves from apex, margins
of valves often with rough fibrous strands, columella not extending to
apex of capsule, with 4-5 hard, sharp, woody ridges; seeds transversely
ellipsoid or suborbicular, narrowly winged all round margin.

Distribution: 7 or possibly 8 species in tropical Africa and Madagascar,
of which 1 introduced in French Guiana.

1. **Khaya senegalensis** (Desr.) A. Juss., in Mem. Mus. Hist. Nat. 19:
   250, t. 21. 1830. – *Swietenia senegalensis* Desr. in Lam., Encycl. 3:
   679. 1791. Type: Senegal, Roussilon (holotype P-LAM).

Tree, up to 20 m tall, bole often crooked, buttresses short or absent. Leaves
up to 25 cm, bright green and shining above, pale grayish green below;
leaflets 4-10 pairs, oblong or oblong-elliptic, 8 × 5 cm, apex rounded or
with a short apiculum. Capsule 4-valved, valves 4.5 × 0.3 cm.

Distribution: Tropical Africa, introduced in the Neotropics, occasionally
planted in gardens in the Guianas; 1 collection studied (FG: 1).

Specimen examined: French Guiana: Cayenne, Centre IRD, Prévost
& Sabatier 4887 (CAY).

7. **MELIA** L., Sp. Pl. 384. 1753.
   Type: *M. azedarach* L.

Monopodial trees or treelets. Bud-scales absent. Indument a mixture of
simple and tufted-stellate hairs. Leaves 2-3-pinnate, leaflet margin serrate,
dentate or crenate. Inflorescences axillary, many-flowered panicles, ultimate
branches often cymose; hermaphrodite and male flowers on same individual
(plant polygamous). Flowers conspicuous; calyx 5-6-lobed to near base,
lobes sometimes imbricate at base; petals pink to lilac-coloured, 5(-6), free,
imbricate; staminal tube narrowly cylindrical, slightly expanded at mouth,
10-12-ribbed, terminated by slender appendages as many as or twice as
many as anthers, anthers 10(-12), hairy or glabrous, inserted on margin of
staminal tube or just inside, alternating with or opposite appendages; nectary
annular or patelliform, free, surrounding base of ovary; ovary 4-8-locular,
locules with 2 superposed ovules, style-head capitate to coroniform, with
4-8 short stigmatic lobes. Fruits fleshy, 3-8-locular drupes, endocarp thick,
bony, hollowed at base and apex, locules 1(-2)-seeded.

Distribution: A complex group of possibly 3 poorly defined species in
the Old World tropics, of which *M. azedarach* is introduced in tropical
America.

1. **Melia azedarach** L., Sp. Pl. 384 (1753). Type: Holland, de Hartecamp, cult. Hort. Cliff. 161.1 (lectotype BM, designated by Mabberley, 1984).

Tree, to 15 m tall (sometimes flowering as a shrub); bark grey-brown, smooth. Leaves 2(-3)-pinnate; petiole and rachis to 40 cm; leaflets opposite or subopposite; petiolule up to 7 mm; leaflet blade lanceolate, 5.5 × 2.5 cm, apex acuminate or subacuminate, base asymmetric, margin deeply crenate or serrate, sparsely puberulous below. Inflorescence a axillary, cymose panicle, many-flowered. Flowers sweetly scented; calyx 0.2 cm, densly stellate puberulous; petals pale lilac, spathulate, 0.8 × 0.3 cm; staminal tube dark purple, 0.7 cm, glabrous outside, hairy inside, appendages 0.1 cm; ovary >0.1 cm diam., 5(-7) locular, style 0.4 cm. Drupe pale yellow, 2 × 1.5 cm.

Distribution: Paleotropics. Introduced in the Neotropics, occasionally grown as an ornamental in the Guianas; 1 collection studied (FG:1).

Specimen examined: French Guiana: Saül, in garden, Vigneron 19 (CAY).

8. **SWIETENIA** Jacq., Enum. Syst. Pl. 4: 20. 1760.
Type: *S. mahagoni* (L.) Jacq. (Cedrela mahagoni L.)

Deciduous trees, monoecious. Shoot apex bearing a cluster of bud-scales. Leaves paripinnate, often with an abortive terminal leaflet, leaflets entire, glabrous. Inflorescences axillary, thyrsoid. Flowers unisexual, greenish-cream coloured; calyx 5-lobed, lobes imbricate, rounded or obtuse; petals (4-)5, much longer than calyx in bud, free, contorted, reflexed in open flower; staminal tube cup-shaped or urceolate, constricted at apex, anthers 8-10, inserted within throat of tube and partially exserted, tube terminated by 8-10 small appendages alternating with anthers; nectary disk in male flowers patelliform, fused to base of staminal tube and forming an annulus around pistillode, in female flowers reduced to a small swelling around base of ovary; ovary (4-)5(-6)-locular, locules with 9-16 ovules, style-head discoid; pistillode more slender, with rudimentary ovules. Fruits erect, strongly woody, oblong to ovoid, septifragal capsules, opening by 5 valves from apex and base simultaneously, columella woody, 5-angled, extending to apex of capsule, with conspicuous seed scars; seeds 9-16 per locule, attached by wing-end to apex of columella.

Distribution: Three species from Florida, the West Indies, southern Mexico and C America to Panama, along the Amazonian slopes of the

Andes to Bolivia, southern and eastern Brazil, but absent from most of the Amazon basin; *S. macrophylla* is occasionally introduced in the Guianas.

1. **Swietenia macrophylla** King, Hooker's Icon. Pl. 16: t. 1550. 1886. Type: India, Calcutta Botanic Garden, King s.n. (holotype K).

Tree, 35-40 m tall. Young branchlets glabrous, slender rich chestnut-brown or reddish-brown, finely lenticellate; bud-scales prominent, brown, ovate or acute. Leaves usually pari-pinnate sometimes imparipinnate with an abortive terminal leaflet, (14-)16-30(-40) cm long, clustered at ends of branchlets; pulvinus swollen, rachis glabrous; leaflets opposite or subopposite, (2-)3-6(-8) pairs; petiolule 5-12 mm long, slender; leaflet blade chartaceous, usually oblong or oblong-lanceolate or ovate-lanceolate, sometimes elliptic-ovate, slightly falcate, (8-)9-13(-18.0) × 3-4(-5.5) cm, apex acute or very short acuminate, base uneven, truncate, rounded or subcordate, glabrous, dark glossy green above; secondary venation prominent adaxially, tertiary veins impressed and obscure. Inflorescence axillary or subterminal, 10-18(-20) cm long with short lateral, spreading branches, generally shorter than leaves, glabrous; pedicels very slender, 1.5-2.5 mm long, glabrous. Flowers unisexual, sexes very similar; calyx 5-lobed, lobes 1-1.5 mm long, broadly rounded, glabrous, margin ciliolate; petals 5, white or greenish white with a pink tinge, free, slightly contorted in bud, oblong to ovate-oblong, (4.5-)5-6 × 2-2.5 mm, glabrous, margin ciliolate; staminal tube 3-4.5 mm, cylindrical, slightly constricted at throat, terminated by 10 acuminate appendages, glabrous inside and out, anthers or antherodes 10, sessile, contained within mouth of tube; nectary bright orange, annular or patelliform, margin crenulate; ovary in female flowers globose, glabrous, (4-)5-locular, locules with (10-)12-14 ovules, style short ca. 1.5 mm long, style-head discoid, with 4-5 glandular stigmatic lobes; pistillode in male flowers very slender, with well-developed locules but rudimentary ovules, style ca. 2 mm long with thinner style-head. Capsule erect, elongate to elongate-ovoid sometimes pear-shaped, 10-15(-22) × 6-8(-10) cm, with a short umbo, greyish-brown, smooth or minutely verrucose, (4-)5-valved, outer valves woody, 6-8 mm thick, inner valves much thinner, mottled brown and white; seeds dark lustrous brown, 7.5-10(-12) cm long including wing.

Distribution: Mexico, C America and northern S America; 1 collection studied (FG: 1).

Specimen examined: French Guiana: Cayenne, Broadway 226 (NY).

9. **TRICHILIA** P. Browne, Civ. Nat. Hist. Jamaica 278. 1756.
Type: *T. hirta* L.

Trees or treelets, usually dioecious, less frequently polygamous. Buds usually naked. Indument usually of simple hairs, less frequently of malpighiaceous, dibrachiate or stellate hairs or peltate scales. Leaves usually pinnate, less frequently 3-foliolate; leaflets sometimes glandular-punctate and -striate. Inflorescences axillary in thyrsoid panicles or rarely fasciculate. Flowers usually unisexual; calyx shallowly to deeply 3-6-lobed or free, aestivation open or imbricate; petals mostly 4-5, free or partially united, imbricate or valvate; filaments completely united to form a staminal tube, or filaments partly free (rarely completely so) and then usually with 2 terminal lobes or appendages, anthers 5-10, hairy or glabrous, inserted between teeth or lobes on margin of staminal tube or apically on filaments; nectary usually a fleshy annulus surrounding base of ovary, or stipitate, cyathiform or patelliform; ovary 2-3(-4)-locular, locules with 1-2 collateral or less frequently superposed ovules, style-head usually capitate. Capsules loculicidal, 2-3(-4)-valved, valves leathery or woody, 1-2-seeded; seeds partly or completely surrounded by a thin fleshy arillode, or rarely with a complete sarcotesta; usually exendospermous; embryo with plano-convex collateral or rarely superposed cotyledons, radicle usually superior and included.

Distribution: A genus of about 14 species in Africa and 85 species in the Neotropics, 17 of which in the Guianas.

KEY TO THE SPECIES

1. Leaflets mostly opposite (except T. surumuensis); petals imbricate or quincuncial, nearly always free; filaments free or partially or completely fused; nectary present or absent 2
   Leaflets mostly alternate; petals valvate, free or united; staminal tube of completely united filaments; nectary absent or represented by an obscure swelling around the base of the ovary 12

2. Stamens free or staminal tube of partially united filaments 3
   Staminal tube of completely united filaments. 10

3. Indument of stellate hairs or peltate scales 4
   Indument not of stellate hairs or peltate scales 5

4. Leaflets alternate; indument minute (x 10); calyx lobes free, imbricate; capsule ellipsoid to globose, 2.5-4 cm long . . . . 4. *T. euneura*
   Leaflets opposite; indument conspicuous; calyx lobes united, with open aestivation; capsule globose to broadly ovoid, 1-1.3 cm long . . . .
   . . . . 6. *T. lepidota* subsp. *leucastera*

5. Ovary locules with 2 superposed ovules. . . . . .10. *T. pallida*
   Ovary locules with 1-2 collateral ovules 6

6. Calyx lobes free, strongly imbricate . . . . .15. *T. septentrionalis*
   Calyx lobes free or united, not or only slightly imbricate at base 7

7. Leaflets densely and coarsely glandular-punctate and-striate. . . . .
   . . . . 3. *T. elegans* subsp. *elegans*
   Leaflets not densely or coarsely glandular-puntate or-striate 8

8. Inflorescence 1-6 cm long, fasciculate or a short panicle 9
   Inflorescence 5-25 cm long, a slender or branched panicle 10

9. Leaves drying pale green, leaflets 3-7, secondary veins 8-12 pairs, petals
   (3-)4, usually appressed puberulous outside, filaments with terminal
   appendages . . . . .10. *T. pallida*
   Leaves not drying pale green, leaflets 7-9, secondary veins 12-15 pairs, petals
   (4-)5, glabrous, filaments usually truncate at the apex . . . . .13. *T. rubra*

10. Leaflets alternate, base of petiole with a pair of minute (0.5-1 mm long)
    caducous scales, capsule dull red, oblong to obovoid. . 17. *T. surumuensis*
    Leaflets opposite, base of petiole without paired scales, capsule yellow to
    orange, broadly ellipsoid, ovoid or globose 11

11. Capsule sericeous, ellipsoid . . . . .8. *T. micrantha*
    Capsule granular-papillose, with some short hairs, ovoid to globose . . . . . .
    . . . . .7. *T. martiana*

12. Leaflets dimorphic or heteromorphic, lowest pairs much reduced and often a
    different shape, sometimes reduced to vestigial scales 13
    Leaflets not dimorphic or heteromorphic 21

13. Petals free or united only at the base 14
    Petals united $1/_3$ - $2/_3$ their length. 18

14. Reduced basal leaflets linear-subulate, 5-7 mm long, or represented by a pair
    of minute caducous scales 15
    At least some of the reduced basal leaflets with an expanded blade 16

15. Leaves 1-3-foliolate, with 1-2 pairs of reduced basal leaflets of 5-7 mm long,
    normal leaflets subcoriaceous, with a prominent tertiary vein reticulum. . .
    . . . . .1. *T. areolata*
    Leaves pinnate, basal leaflets represented by a pair of caducous scales of 1-2
    mm long, normal leaflets chartaceous, with obscure tertiary venation . . . .
    . . . . .12. *T. quadrijuga* subsp. *quadrijuga*

16. Mature capsule dark green, glabrous, nearly always verrucose . . . .
    . . . . .11. *T. pleeana*
    Mature capsule greyish, greyish-green, dark green or reddish, sericeous,
    smooth 17

17   Leaflets 7.5-10.5 cm long, base usually asymmetric; petals (2-)2.5-3.5 mm long; capsule 1.2-1.4 cm long; seed completely surrounded by a thin arillode . . . . 5. *T. lecointei*
      Leaflets 7.8-27 cm long, base usually regular; petals 3.5-4.5(-5) mm long; capsule 2-3.5 cm long; seed only at apex with a strongly developed arillode .14. *T. schomburgkii* subsp. *schomburgkii*

18   Reduced basal leaflets represented by a pair of subulate scales of 1-2 mm long. . . . . .12. *T. quadrijuga* subsp. *quadrijuga*
      Reduced basal leaflets usually more than 1 pair, always with an expanded blade 19

19   Petals 1.7-2.5 mm long . . . . .9. *T. micropetala*
      Petals 3-4.5 mm long 20

20   Mature capsule glabrous, dark green, nearly always verrucose . . . .
      . . . . 11. *T. pleeana*
      Mature capsule puberulous, not green, smooth . . . .
      . . . . .14. *T. schomburgkii* subsp. *schomburgkii*

21   Leaflets 7-11, secondary veins mostly 16-18 pairs, shallowly ascending; petals appressed puberulous . . . . .2. *T. cipo*
      Leaflets 5-7(-8), secondary veins mostly 11-13, strongly arcuate ascending; petals subglabrous . . . . .16. *T. surinamensis*

1. **Trichilia areolata** T.D. Penn., Fl. Neotrop. 28: 208. 1981. Type: Brazil, Amazonas, Manaus, Coelho & Mello 3007 (holotype INPA, isotype FHO).   – Fig. 11

Tree, to 17 m tall; bark pale brown, with shallow horizontal and vertical fissures, with slight orange exudate and spicy meliaceous scent. Young shoots slender, 2-4 mm diam., coarsely pubescent with crisped, spreading and erect pale hairs, becoming glabrous, rough, greyish-brown with numerous pale lenticels. Bud-scales absent. Leaves 1-3-foliolate; petiole of 1-foliolate leaves 1.3-2.5 cm long, petiole of 3-foliolate leaves ca. 1 cm long, semiterete, sparsely and coarsely pubescent; leaflets opposite, dimorphic, 1-3, with 1-2 pairs of vestigial, linear-subulate "pseudostipules" at base of petiole, these 0.5-0.7 cm long; petiolule of lateral leaflets 3-4 mm long; lateral leaflet blade subcoriaceous, elliptic, ca. 5 × 3 cm, apex acute, base asymmetric, acute/obtuse, terminal leaflet and 1-foliolate leaflet blade elliptic or oblanceolate, 10 × 4.2-15.5 × 7.2 cm, apex acuminate to obtusely cuspidate, base cuneate or acute, glandular-punctate and -striate, adaxial surface densely pubescent along midrib, abaxial surface coarsely pubescent along midrib and veins; venation eucamptodromous to brochidodromous in upper half, midrib prominent on adaxial surface, secondaries 7-15 pairs, ascending, arcuate or +/-straight, parallel or slightly convergent, intersecondaries obscure or absent, tertiaries forming a conspicuous reticulum. Inflorescence axillary, 10-20 cm long, an irregularly branched slender panicle, sparsely and coarsely pubescent; pedicels 0.5-1.5 mm long. Flowers with calyx cyathiform, 1-1.5 mm long, 5-toothed, teeth triangular, $^1/_3$-$^1/_2$ the length of calyx, subglabrous; petals whitish, 5, free, valvate, narrowly lanceolate, 3-3.5 × 0.7-1 mm, apex acute, hooded, strigulose outside, glabrous inside; staminal tube cyathiform, 1.5-2 × ca. 1.5 mm, filaments completely fused, alternate filaments slightly longer, bearing 2 slender apical appendages about $^1/_2$ the length of anthers, glabrous outside, crisped pubescent inside, anthers 9-10, 0.6-0.7 mm long, glabrous; nectary absent; ovary ovoid, 0.75-1 mm long, densely strigose, 2-3-locular, locules with 2 collateral ovules, style glabrous, style-head capitate, equalling anthers. Capsule ellipsoid, 2.3-2.5 × 1.3-1.5 cm, apex and base obtuse to rounded, smooth, densely tomentose, 3-valved, pericarp ca. 1.5 mm thick; seeds black, solitary, obovoid, ca. 2 × 1 cm (including arillode), apex truncate, base obtuse, arillode orange, developed at apex of seed and along adaxial surface only, leaving abaxial face exposed.

Distribution: Southern Suriname to central and eastern Amazonian Brazil; in lowland rain forest on both periodically flooded and non-flooded areas, below 200 m alt.; 4 collections studied (SU: 4).

Specimens examined: Suriname: Sipaliwini, vicinity of Ulemari R., 150 km upstream from confluence with the Litani R., Evans & Peckham 2894 (K), Hammel et al. 21388 (K), 21527 (K), 21569 (K).

Phenology: Flowering in November and December and fruiting March to April.

2. **Trichilia cipo** (A. Juss.) C. DC. in Mart., Fl. Bras. 11(1): 214. 1878. – *Moschoxylum cipo* A. Juss., Bull. Sci. Nat. Géol. 23: 239. 1830. Type: French Guiana, banks of Kourou R., Collector unknown s.n. (holotype P (ex herb. L.C. Richard)).

Tree, to 12 m tall, dioecious; bark smooth, greyish to mid brown. Young branches appressed puberulous (most trichomes basifixed) at first, becoming glabrous, brown to greyish-white, lenticellate. Leaves imparipinnate or pinnate with one leaflet of ultimate pair orientated to simulate a terminal leaflet, 7-11-foliolate; petiole 2.5-7 cm long, semiterete or narrowly winged at base, rachis 2-22 cm long, semiterete becoming terete in upper half, minutely puberulous, hairs erect or appressed; leaflets alternate or less frequently opposite; petiolule 1.5-3 mm long; leaflet

116

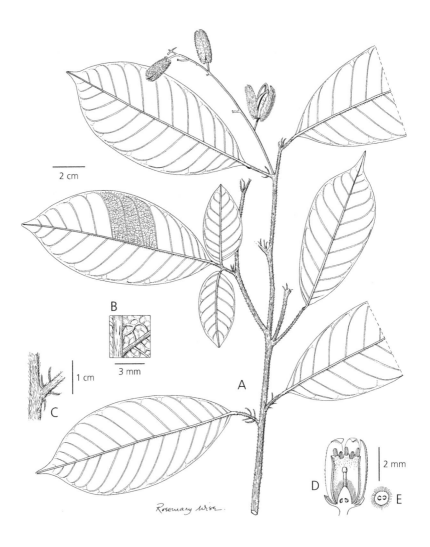

Fig. 11. *Trichilia areolata* T.D. Penn.: A, habit with fruit; B, detail of leaflet abaxial surface; C, enlargment of stipules; D, flower, longitudinal section; E, ovary, cross section. (A-C, Hammel & Koemar 21388; D, Coelho & Mello 3007).

blade subcoriaceous, oblong to elliptic, 10-16 × 3.5-6 cm, basal leaflets smaller, apex acuminate to obtusely cuspidate, base cuneate to attenuate, adaxial surface glabrous or with midrib puberulous, abaxial surface with a few scattered appressed hairs or glabrous, sometimes glandular-punctate and -striate; venation eucamptodromous or rarely brochidodromous, midrib usually prominent adaxially, secondaries (13-)16-18 pairs, usually rather shallowly ascending, arcuate, slightly convergent to +/- parallel, intersecondaries short, usually rather obscure. Inflorescence axillary, narrowly pyramidal, 10-40 cm long, rather laxly-branched, few- to many-flowered thyrse, minutely puberulous; pedicels 0.5-1(-2) mm long. Flowers unisexual, female flowers usually fewer and somewhat larger than male; calyx patelliform to shallowly cyathiform, 0.5-1.5 mm long, lobes 4-5, aestivation open, broadly ovate, triangular, acute or obtuse, sparsely appressed puberulous; petals cream-coloured, fused $1/3$-$2/3$ their length, valvate, ovate to lanceolate, (3-)4-5, 2-2.5 × 0.7-1.5 mm, acute, appressed puberulous outside, glabrous inside; staminal tube urceolate, filaments completely fused, 1-1.7 mm long and broad, margin with 1(-2) lanceolate or subulate lobes alternating with anthers, glabrous, anthers 7-8, 0.4-0.7 mm long, glabrous; antherodes slender, not dehiscent, without pollen; nectary absent; ovary ovoid, densely puberulous to strigulose, (2-)3-locular, locules with 2 collateral ovules, style glabrous, at least at apex, style-head capitate rarely minutely apiculate, held below or at level of anthers; pistillode smaller, +/- conical, containing well-developed, nonfunctional ovules. Capsule greyish, ellipsoid to botuliform, 1.3-2 × 0.6-1.1 cm, smooth, moderately to densely appressed puberulous, 3(-4)-valved, pericarp ca. 0.5 mm thick, endocarp thin, cartilaginous; seeds 2, collateral in each fruit, 1-1.5 cm long, arillode dark red, fleshy, strongly developed only at apex of seed, seed coat membraneous.

Distribution: Guyana, French Guiana, throughout central and western Amazonia, and state of Maranhão, Brazil; lowland rain forest, usually recorded from riverbanks and the edges of black water creeks, alt. 50-600 m; 23 collections studied (GU: 14; SU: 1; FG: 4).

Selected specimens: Guyana: without locality, Ro. Schomburgk ser. I, 668 (K, US); Rupununi, Kuyuwini landing, Kuyuwini R., Jansen-Jacobs et al. 3082 (K). Suriname: Tumuc Humac Mts., W of Talouakem Peak, Acevedo-Rodriguez et al. 6126 (K). French Guiana: Inselberg Talouakem, Massif des Tumuc-Humac, de Granville 12122 (K); Marouini, environs d'Antecume Pata, Cremers 5000 (K); Sinnamary R., Oldeman B-1233 (K); Crique Courouaie, above Deux Fourches, Oldeman B-2697 (CAY).

Phenology: Flowering from August to December and fruiting in March.

3. **Trichilia elegans** A. Juss. in A. St.-Hil., Fl. Bras. Mer. 2: 79, t. 98. 1829. Type: Brazil, "Prov. Sao Paulo", Saint-Hilaire s.n. (annotated 'Cat 2, 1236'), (holotype P, isotypes P).

In the Guianas only: subsp. **elegans**   – Fig. 12

Tree, to 20 m tall, dioecious; bark dark brown to greyish with prominent pale lenticels. Young branches slender, at first pubescent to densely golden villose usually soon glabrous. Bud-scales absent. Leaves imparipinnate, 5-7(-9)-foliolate; petiole 2-7 cm long; rachis 4-11 cm long, both semiterete, nearly always flattened and slightly broader below each pair of leaflets, usually puberulous or glabrous; leaflets opposite or subopposite; petiolule 0-2(-3) mm long, often appearing longer due to long decurrent base of leaflet, petiolule of terminal leaflets often longer, to 1.5 cm; leaflet blade chartaceous to subcoriaceous, usually elliptic, less frequently oblanceolate, 3.5-17.5 × 1-6.2 cm, apex usually attenuate or acuminate, base usually attenuate and often long decurrent, sometimes slightly asymmetric, terminal leaflet often larger, leaflet blade gradually reduced in size towards base of leaf, margin usually revolute, adaxial surface glabrous, abaxial surface nearly always with prominent tufts of hairs (domatia) in frequently hollowed out vein axils, otherwise glabrous, granular red papillae sometimes present, densely pellucid-punctate and -striate; venation eucamptodromous or less frequently brochidodromous, midrib flat or slightly prominent on adaxial surface, secondaries (6-)9-13 pairs, shallowly to steeply ascending, parallel, straight to arcuate, intersecondaries prominent. Inflorescence axillary, usually slender to broadly pyramidal, 3-10 cm long, laxly- to rather densely-flowered, thyrse, rarely racemose, sparsely pubescent or glabrous; pedicels (0-)0.3-1 mm long. Flowers white or greenish white, unisexual; calyx cyathiform, 1-1.5 mm long, lobes 5, usually free, quincuncial, triangular, acute, usually ciliate and sparsely puberulous, persistent in fruit; petals 5, free, imbricate or rarely quincuncial, oblong-lanceolate or elliptic, 2-2.5 × 0.75-1.25 mm, apex rounded, acute, or obtuse, usually glabrous, rarely ciliate; staminal tube cyathiform or urceolate, 0.7-1.2 × 0.7-1.2 mm, filaments completely fused and margin with or without 10 short acute appendages alternating with anthers, outside of tube with few minute crisped hairs in upper half or glabrous, inside barbate to puberulous in throat, less frequently glabrous, anthers 10, 0.7-1 mm long, usually glabrous rarely sparsely pubescent; antherodes slender, indehiscent, without pollen; nectary in male flower a fat annulus enveloping pistillode, in female reduced to a small swelling around base of enlarged ovary, glabrous; ovary ovoid or conical, glabrous, (2-)3-locular, locules with 2 collateral ovules, style short, stout, glabrous, surmounted by a capitate or discoid style-head with 3 stigmatic lobes, below anthers; pistillode much smaller, with small non-functional ovules. Capsule maroon or wine-red, usually ellipsoid, 0.7-1 × 0.5-0.8 cm, apex acute, apiculate, base usually truncate, smooth, densely granular papillose, with or without some puberulous or pubescent indument, 3-valved, pericarp 0.5 mm thick, outer part fleshy, endocarp thick, cartilaginous; seeds 1-3 in each fruit, ellipsoid or shaped like segment of an orange, 0.6-0.8 × 0.4-0.7 cm, provided with a thick fleshy arillode which develops only around upper $1/2$ and adaxial surface of seed, and also round aborted ovules, arillode orange-red, free except along a thin line of attachment from micropyle along short ventral raphe, seed coat tough, shining.

Distribution: Most abundant in southern Brazil, Paraguay and northern Argentina, but extending northwards through eastern and central Bolivia, and along the eastern slopes of the Andes through Peru and Colombia, Venezuela and Guyana; 9 collections studied (GU:4).

Specimens examined: Guyana: Kanuku Mts., Maipaima, camp 3 on Tsikoma Cr., Jansen-Jacobs *et al.* 1006 (K, U); U. Takutu-U. Essequibo Reg., E slope of Shiriri Mt., Peterson *et al.* 7722 (K, US); Rupununi Distr., NE slope of Mt. Shiriri, Jansen-Jacobs *et al.* 4160 (K, U, US); Marudi Mts., along trail from Norman Mines, Stoffers *et al.* 282 (U).

Phenology: Flowers recorded in November, fruits in June.

Note: The other subsp. *richardiana* (A. Juss.) T.D. Penn. is occurring in Brazil.

4. **Trichilia euneura** C. DC. in A. DC. & C. DC., Monogr. Phan. 1: 673. 1878. Type: French Guiana, Cayenne, Leprieur s.n. (holotype G-DC, isotypes G, K, P). – *Trichilia stelligera* Radlk., Repert. Spec. Nov. Regni Veg. 9: 372. 1911. Type: Suriname, Tresling 281 (holotype M, isotype U).

Tree, to 20 m tall, dioecious, bole in larger trees slightly fluted at base; bark greyish, lenticellate, scaling in long slender pieces. Young branches minutely stellate-puberulous becoming greyish-brown and glabrous, usually without lenticels. Bud-scales absent. Leaves imparipinnate or pinnate with one leaflet of ultimate pair orientated to simulate a terminal leaflet, 4-7 foliolate, 10-22 cm long; petiole 2.5-8 cm long, rachis 4-21 cm long, semiterete, minutely stellate-puberulous to subglabrous; leaflets alternate or rarely subopposite; petiolule 3-10 mm long; leaflet blade coriaceous, oblong to broad elliptic, less frequently oblanceolate, (8-)13-20 × (3.5-)6-10 cm, apex obtusely cuspidate to narrowly attenuate

120

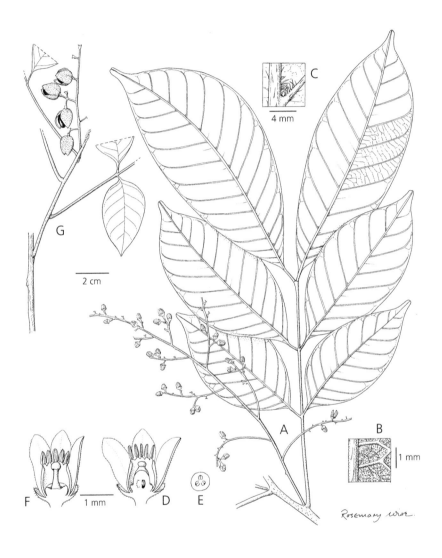

Fig. 12. *Trichilia elegans* A. Juss.: A, habit; B, enlargement of glandular lines and dots on leaflet; C, axillary hair tufts; D, E, female flower; F, male flower; G, infructescence. (A-E, Jansen-Jacobs *et al.* 1006; F, Stoffers *et al.* 282; G, Jansen-Jacobs 4160).

or acuminate, base narrowly or broadly cuneate or attenuate, both surfaces sparsely to moderately minutely stellate or peltate-stellate or subglabrous, glandular-punctate and -striate; venation eucamptodromous, midrib slightly sunken to slightly prominent, secondaries (6-)9-20 pairs, steeply ascending, usually straight and parallel, less frequently slightly arcuate and convergent, intersecondaries obscure or absent, tertiaries oblique and parallel, often closely spaced, moderately prominent on both sides. Inflorescence axillary, 2.5-7 cm long, a densely-flowered thyrse, stellate puberulous; pedicels 0.5-1.5 mm long. Flowers unisexual; calyx shallowly cyathiform, 1-2 mm long; lobes (4-)5, free, imbricate, broadly ovate to orbicular, stellate-puberulous, often ciliate; petals white, cream or yellowish green, 5, free, imbricate, oblong, elliptic or lanceolate, 4-5 × 1-2.5 mm, apex rounded, obtuse or acute, stellate-puberulous outside, glabrous inside; staminal tube urceolate or short cylindrical, 3-4 × 1.5-2.3 mm long and broad, filaments fused $1/3$-$1/2$ their length, terminated by 2 acute appendages from $1/3$ to equalling anthers in length, indument outside varying from uniformly stiff pubescent (simple hairs) to glabrous, inside usually with a mixture of stellate and simple hairs, or less frequently glabrous, anthers 8 or 10, 0.4-1 mm long, usually with a few stiff hairs on dorsal side, rarely uniformly pubescent; antherodes slender, without pollen, not dehiscent; nectary a small annular pubescent swelling around base of ovary or absent; ovary ovoid, densely stellate pubescent, 3-locular, locules with 2 collateral ovules, style densely stellate pubescent, style-head clavate, capitate or rarely discoid, below anthers; pistillode flattened-conical, containing small, non-functional ovules. Capsule yellow to bright orange, ellipsoid to obovoid, 2.7-3.8 × 0.7-1.6 cm, smooth or minutely tuberculate, densely stellate puberulous, 3-valved, valves remaining erect, apex acute to truncate, base acute, pericarp 1.5-2 mm thick, endocarp thin cartilaginous; seeds orange, solitary, ellipsoid or oblong, 2.5-3.2 × 0.9-1.1 cm, acute at base and apex, completely surrounded by a fleshy sarcotesta (the only species of *Trichilia* that has this).

Distribution: Western and central Amazonia to northern Venezuela, Suriname, French Guiana and state of Amapá, Brazil; non-flooded lowland forest, alt. 170-300 m; 25 collections studied (SU: 2; FG: 20).

Selected specimens: Suriname: Litani R., Acevedo-Rodriguez *et al.* 5748 (K); Tumuc-Humac Mts., Talouakem, upper Litani R., Acevedo-Rodriguez *et al.* 5912 (K). French Guiana: Approuague R., entre le saut le crique Couata, Oldeman T-217 (K); Saül, Eaux Claires, Prance 30677 (K); Camopi R., above the confluence with Tamouri R., Lescure 142 (K); Borne Frontier No 1, Molino & Sabatier 2330 (CAY).

Phenology: Flowering and fruiting from June to September.

122

5. **Trichilia lecointei** Ducke, Arch. Jard. Bot. Rio de Janeiro 3: 191. 1922. Type: Brazil, Pará, Obidos, Le Cointe RB16799 (lectotype G, designated by Pennington 1981, isolectotypes BM, K, P, US).

Tree, to 30 m tall. Young branches puberulous to densely short pubescent (trichomes basifixed) becoming glabrous, greyish, rough and fissured. Leaves imparipinnate or pinnate with one leaflet of ultimate pair orientated to simulate a terminal leaflet, 6-8-foliolate (excluding basal 2-3 pairs of reduced leaflets), 11-16 cm long; petiole 3-5.1 cm long, semiterete, rachis usually +/- terete sometimes canaliculate, puberulous; leaflets alternate to opposite, sometimes dimorphic or more often with a gradual transition between normal leaflets and the much reduced basal ones; petiolule 2-3(-5) mm long; leaflet blade subcoriaceous, oblong or elliptic, 7.5-10.5 × (2.5-)3.3-4.5(4.0) cm, apex acuminate, base usually asymmetric and acute to obtuse, adaxial surface glabrous or with midrib sparsely puberulous, abaxial surface glabrous or rarely midrib puberulous, sometimes glandular-punctate and -striate, basal reduced leaflets usually persistent, ovate to reniform, 0.3-1.5(-1.7) cm long, with a strongly oblique base, petiolulate; venation eucamptodromous, midrib flat or slightly prominent, secondaries 12-13 pairs, ascending, slightly arcuate or straight, parallel or slightly convergent; intersecondaries obscure. Inflorescence axillary, 10-20 cm long, a rather densely-flowered, broadly pyramidal panicle, puberulous; pedicels 1-1.5 mm long. Flowers bisexual; calyx patelliform, 0.5-1 mm long, lobes 4-5, ovate or broadly triangular, ca. 1/2 length of calyx, acute to obtuse, sparsely puberulous; petals whitish, 4-5, free or fused at base, valvate, reflexing, lanceolate, (2-)2.5-3.5 × 1-1.5 mm, acute, sparsely appressed puberulous outside, glabrous inside; staminal tube urceolate or cyathiform, filaments completely fused, 1-2 × 1.5-2 mm, margin with 7-10 triangular, acute or subulate lobes alternating with anthers, glabrous or occasionally with scattered long hairs inside throat, anthers 7-10, 0.7-0.8 mm long, glabrous; nectary absent; ovary conical, densely pubescent, 3-locular, locules with 2 collateral ovules, style glabrous at least at apex, style-head capitate, level with base of anthers. Capsule ovoid, 1.2-1.4 cm long, smooth, minutely sericeous, 3-valved, pericarp ca. 0.5 mm thick, endocarp thin, cartilaginous; seeds bright red, solitary, ca. 0.8 × 0.4 cm, completely surrounded by a thin fleshy arillode which also develops around aborted ovules, arillode free except for thin line of attachment from micropyle to raphe, seed coat thin.

Distribution: Southern Venezuela, Guyana and Brazil (Pará and NW Amazon); in non-flooded forest, from sea level to 600-800 m alt.; 6 collections studied (GU: 1).

Specimen examined: Guyana: Akarai Mts., between drainage of

Mapuera R. (Trombetas tributary) and Shodikar Cr. (Essequibo tributary), A.C. Smith 2965 (K, NY, U).

Phenology: Flowering from August to October and fruiting in January.

6. **Trichilia lepidota** Mart., Flora 22. Beibl. 1: 54. 1839. Type: Brazil, Bahia, near Ilhéus, Martius 393 (holotype M, isotypes BM, BR, F, G, GH, K, NY, P).

In the Guianas only: subsp. **leucastera** (Sandwith) T.D. Penn., Fl. Neotrop. 28: 40. 1981. – *Trichilia leucastera* Sandwith, Kew Bull. 328. 1933. Type: Guyana, Simuni Cr., Rupununi R., Davis in Forest Dept. 2140 (holotype K).

Tree, 15-25 m tall, dioecious, trunk slightly fluted at base. Young branches densely peltate-lepidote to stellate-tomentose, indument usually long persistent, greyish-brown, rough, without lenticels. Leaves imparipinnate, 5-9-foliolate; petiole 6.5-10 cm long, rachis 4.5-15 cm long, petiole and rachis usually semiterete, stellate-pubescent; leaflets opposite, chartaceous to subcoriaceous; petiolule 2-4(-14) mm long; leaflet blade oblong or oblanceolate, less frequently elliptic, 8-20 × 3-8 cm, apex broadly attenuate to narrowly acuminate, base narrowly to broadly attenuate or cuneate, sometimes slightly oblique, especially lower leaflets, upper surface usually glabrous, lower surface moderately or densely stellate-pubescent, denser on midrib and veins; venation eucamptodromous, midrib on upper surface nearly always sunken, secondaries 12-16 pairs, ascending, usually straight and parallel, intersecondaries and tertiaries usually obscure. Inflorescence axillary, (8-)11-20 cm long, a slender to broadly pyramidal, laxly- to densely-flowered thyrse, indument of dense stellate hairs; pedicels 1-2 mm long. Flowers unisexual; calyx patelliform or shallowly cyathiform, 1-1.5 mm long, with 5 shallow, usually narrowly attenuate lobes, 1/4-1/3 length of calyx, indument of dense stellate hairs, calyx persistent in fruit; petals (4-)5, free, slightly imbricate, lanceolate to oblong, 3-4 × 1-1.5 mm, apex acute, outside densely appressed stellate-puberulous, inside glabrous; filaments 1-2(-3) mm long, +/- free, those alternate usually slightly shorter, filament apex with 2 short acute lobes to 1/3 length of anther, glabrous, anthers 8-10, 0.4-0.6 mm long, glabrous; antherodes narrow, shrunken, not dehiscent, without pollen; nectary represented by a small annulus around base of ovary or absent, stellate-pubescent; ovary ovoid, densely stellate-tomentose, 3(-4)-locular, locules with 2 collateral ovules, style short, usually glabrous at apex, style-head clavate to capitate, usually obscurely 3-lobed; pistillode flattened-conical or small ovoid, containing small non-functional ovules. Capsule drying pale yellowish-brown, globose to broadly ovoid, 1-1.3 cm long, apex rounded, base with a short stout stipe of 1-2 mm long, smooth, densely stellate-pubescent, 3(-4)-valved, valves opening widely but not reflexing, pericarp 1-1.5 mm thick, endocarp very tough-cartilaginous; seeds 1-2, collateral in each valve, 0.5-0.7 × 0.3-0.4 cm, with a small, thick, fleshy adaxial arillode extending from apex to base of seed, seed coat thin, hard, smooth, shining brown.

Distribution: Eastern Venezuela (Bolìvar), the Guianas and Brazil (Maranhão); lowland forest up to 700 m alt.; 6 collections studied (GU: 1; SU: 3; FG: 2).

Selected specimens: Suriname: Lely Mts., SW plateau, Lindeman & Stoffers *et al.* 750 (U); Nickerie Distr., area of Kabalebo Dam project, Lindeman & de Roon *et al.* 830 (K). French Guiana: Forêt de Cr. Plomb, Sabatier *et al.* 4756 (K); Station des Nouragues, Bassin de l'Arataye, Sabatier *et al.* 1827 (CAY, K).

Vernacular names: Guyana: ulu (Arawak). Suriname: melisali.

Phenology: Apparently with two flowering periods, the first in January and February, the second in July and August.

Note: The typical subsp. *lepidota* and subsp. *schumanniana* (Harms) T.D. Penn., are both confined to the Atlantic coastal forests of Brazil.

7. **Trichilia martiana** C. DC. in Mart., Fl. Bras. 11(1): 205. 1878. Type: Brazil, near Cajú, Riedel s.n. (holotype LE, isotypes G-DC, K, LE, P).

Trichilia fuscescens Radlk., Sitzungsber. Math.-Phys. Cl. Königl. Bayer. Akad. Wiss. München 9: 641. 1879. Type: Suriname, Lawa R., Kappler 2130 (holotype M, isotypes GOET, P).

Tree, to 14(-35) m tall (often flowering as a small treelet of 2-3 m), dioecious; bark smooth, pale greyish. Young branches puberulous to pubescent or glabrous, mid- to pale brown or greyish, with small lenticels. Leaves imparipinnate, 7-9-foliolate; petiole 3-11 cm long, rachis 3-17 cm long, both terete or semiterete, pubescent or glabrous; leaflets opposite or subopposite; petiolule (1-)2-5(-9) mm long, petiolule of terminal leaflet usually much longer than laterals; leaflet blade chartaceous, oblanceolate or cuneiform, 10-14 × 3-7 cm, apex usually attenuate, acuminate or obtusely cuspidate, rarely rounded, base acute or cuneate, adaxial surface glabrous or midrib puberulous to pubescent, abaxial surface pubescent or glabrous, with or without granular red papillae, usually finely glandular-punctate and -striate; venation eucamptodromous, midrib flat or slightly sunken on adaxial surface, secondaries 12-14 pairs, slightly arcuate, parallel. Inflorescence in axils of new leaves, 5-20 cm long, paniculate, usually a slender panicle, flowers often in rather dense umbellate fascicles on short lateral branches, sparsely and minutely puberulous to villose; pedicels 1-2(-3) mm long. Flowers unisexual, fragrant; calyx patelliform to shallowly cyathiform, 1-1.5 mm long, lobes (4)5(-6), aestivation open, triangular, 1/2-3/4 length of calyx, acute, puberulous, appressed puberulous or pubescent; petals creamy to greenish white, (4)5(-6), free, imbricate, oblong to lanceolate or rarely elliptic, 2.5-4 × 1-1.5 mm, apex acute, usually appressed puberulous and papillose or appressed pubescent outside, sometimes subglabrous, papillose or glabrous inside; staminal tube cyathiform, 1.5-2 × 1.5-2.5 mm, filaments usually fused 1/4-1/2 their length, apex rounded or truncate or terminated by 2 short acute lobes, sparsely coarse pubescent to glabrous outside, inside glabrous at base then barbate to throat, anthers 8-10, 0.4-0.6 mm long, often prolonged slightly into a short point, sparsely hairy; antherodes narrower, not dehiscing, without pollen; nectary in male flowers a swollen annulus surrounding base of vestigal ovary and fused to base of staminal tube, reduced or absent in female flowers, nearly always pubescent; ovary ovoid, densely pubescent, 2-3 locular, locules with 2 collateral ovules, style stout, short, pubescent, style-head capitate or clavate; pistillode conical, with or without vestigial ovules. Capsule yellow to orange, ovoid or globose, 0.9-1.3 cm long, with a truncate base, 2-3-valved, valves reflexing, smooth, drying trigonous and a characteristic pale or dark brown, densely granular papillose intermixed with sparse or less frequently moderately dense short hairs, pericarp 0.3-0.5 mm thick, leathery, endocarp thin, cartilaginous; seeds 2, collateral in each valve, 0.7-1.1 cm long, surrounded by a thin fleshy orange or red arillode, basal part of seed sometimes exposed; arillode thickened around apex of seed and along adaxial side, seed coat thin and soft, containing numerous large pale fat bodies.

Distribution: From the Pacific and Caribbean slopes of SE Mexico through C America and northern S America, from the Guianas through southeastern Brazil; lowland rain forest, up to 600 m alt. in the Guianas; 23 collections studied (GU: 1; SU: 12; FG: 3).

Selected specimens: Guyana: Basin of Kuyuwini R. (Essequibo tributary), about 150 miles from mouth, A.C. Smith 2602 (K, NY). Suriname: Wayombo, Stahel 365 (K, NY); bank of Tanjimamma Cr. (tributary Coppename R.), Mennega 480 (NY); Tumuc Humac Mts., Talouakem, U. Litani R., Acevedo-Rodriguez 5907 (K, NY). French Guiana: Maroni, Rech s.n. (U, NY, P, VEN); Massif des Emerillons, Molino & Sabatier 2483 (CAY); Oyapock R., above Saut Moutouci, Oldeman T-734 (K, US).

Vernacular name: Suriname: karababalli djamaro.

Phenology: In the Guianas flowering in August and November.

8. **Trichilia micrantha** Benth., Hooker's J. Bot. Kew Gard. Misc. 3: 369. 1851. Type: Brazil, Amazonas, Barra do Rio Negro [now Manaus], Spruce 1417 (holotype K, isotypes BM, C, G, G-DC, GH, GOET, M, NY, OXF, P).

Trichilia roraimana C. DC. in A. DC. & C. DC., Monogr. Phan. 1: 670. 1878. Type: Guyana, Roraima, Ro. Schomburgk ser. II, 1005 (= Ri. Schomburgk 1735) (holotype G-DC, isotypes G, K, NY, P).
Trichilia acariaeantha Harms, Notizbl. Bot. Gart. Berlin-Dahlem 9: 431. 1925. Type: Suriname, Brownsberg, BW 6453 (holotype U).

Tree, to 20 m tall (sometimes flowering when less than 5 m), monoecious or dioecious. Young branches appressed puberulous to sericeous at first soon glabrous, bark mid-brown to greyish-white, with conspicuous pale lenticels. Bud-scales absent. Leaves imparipinnate, (5-)7-9 foliolate; petiole 3.5-9 cm long, semiterete or sometimes narrowly winged, undersurface often with flat, ovate extra-floral nectaries, rachis 7-18 cm long, semiterete or flattened, puberulous or glabrous; leaflets opposite or subopposite, petiolule 1-6 mm long; leaflet blade chartaceous to subcoriaceous, oblong, elliptic or oblanceolate, rarely oblong-lanceolate, (10-)12-17(-19) × 3-6(-7) cm, apex usually acuminate, less frequently obtusely cuspidate, base usually attenuate, less frequently cuneate or obtuse, sometimes long decurrent, adaxial surface puberulous to glabrous, abaxial surface usually glabrous, rarely puberulous, less frequently puberulous on midrib, sometimes glandular-punctate and -striate; venation eucamptodromous, upper midrib nearly always raised, secondaries 11-20 pairs, usually rather shallowly ascending, arcuate, slightly convergent or parallel, intersecondaries prominent, short or long. Inflorescence usually clustered around shoot apex, in axils of caducous undeveloped leaves or rarely solitary in axils of normal leaves, 8-25 cm long, a slender or broad pyramidal panicle with widely spreading lateral branches up to 6 cm long, flowers in rather dense clusters; pedicels 0.5-1.2 mm long. Flowers yellowish-green or whitish, sweet scented, unisexual or bisexual, puberulous; calyx patelliform, 0.5-0.7 mm long, lobes 5(-6) free, open or slightly imbricate, ovate to suborbicular, ciliate, rarely puberulous; petals (4-)5(-6), free, imbricate or rarely quincuncial, oblong, ovate or less frequently elliptic, 2-2.5mm × 1-1.5 mm, apex acute, outside papillose, sparsely appressed puberulous, inside glabrous, sometimes sparsely ciliate; staminal tube cyathiform or urceolate, 1-1.5 × 0.7-2 mm, filaments fused at base $1/4$-$1/3$ their length, apex rounded or with 2 short acute to subulate appendages, outside glabrous below but with sparse to dense crisped hairs in upper half, inside glabrous below but densely barbate in throat, anthers (7-)10(-11), 0.3-0.5 mm long, sparsely pubescent or glabrous; antherodes shrunken, not dehiscent, without pollen; nectary a thick fleshy, glabrous, annulus surrounding base of ovary, sometimes on a short stout stipe; ovary ovoid, glabrous, 3(-4)-locular, locules 1-ovulate, style glabrous, style-head minute capitate; pistillode reduced, immersed in fleshy nectary, containing smaller non-functional ovules. Capsule yellowish, usually broadly ellipsoid, less frequently ovoid, 1-2.3 × 0.7-1.3 cm, apex obtuse, smooth, sericeous, tomentose, or glabrous, 3-valved, valves opening widely but not reflexed, pericarp 0.5-0.75 mm thick, endocarp cartilaginous; seeds 1-3, 0.8-1.5 cm long, completely surrounded by a thin fleshy 3-partite, bright red arillode, arillode free except for attachment along short adaxial raphe, seedcoat thin, tough.

Distribution: From Trinidad and northern Venezuela, the Guianas, across central Amazonia to Amazonian Colombia, Peru and Bolivia; non-flooded lowland forest, 200-500 m alt.; 71 collections studied (GU: 7; SU: 30: FG: 14).

Selected specimens: Guyana: Potaro-Siparuni Reg., Kaieteur F. Nat. Park, Hahn et al. 4772 (K); Moraballi Cr., Essequibo R., Fanshawe 1287 (K, NY). Suriname: Jodensavanne-Mapane Cr., Suriname R., Schulz 7203 (K, NY, US); Sipaliwini, Ulemari R., Evans 2916 (K). French Guiana: Saül, route de Bélizon, Mori et al. 24820 (K); Bassin du Maroni, Inini R., Sabatier et al. 3333 (K); Piste de St. Elie, Molino 986 (CAY).

Vernacular names: Guyana: subuleroballi (Arawak). Suriname: melisalie, tweede melisalie, sali, witte sali, sorosali.

Phenology: Flowering in April and May and fruiting from May to August.

9. **Trichilia micropetala** T.D. Penn., Fl. Neotrop. 28: 172, fig. 32. 1981. Type: Brazil, Amapá, R. Araguari, Pires, Rodrigues & Irvine 51334 (holotype INPA, isotype FHO).

Tree, to 20 m tall, dioecious. Young shoots puberulous (trichomes basifixed), soon glabrous, brown, lenticellate. Leaves 3-foliolate or pinnate, 1-5 cm long; petiole and rachis semiterete, puberulous; leaflets opposite or alternate, dimorphic, 3-5 with 1-2 pairs of greatly·reduced leaflets at base of petiole; petiolule 2-4 mm long; normal leaflet blade usually elliptic, 6 × 2.2-10 × 4.6 cm, apex acuminate, base narrowly attenuate (terminal leaflet up to 14 cm long), subglabrous, sometimes glandular-punctate and -striate;

venation eucamptodromous, midrib prominent adaxially, secondary veins 8-12 pairs, arcuate-ascending, convergent, intersecondaries short, tertiaries +/-oblique; reduced basal leaflet blade lanceolate or ovate, 0.2-1.6 cm long, with a cordate base or reniform, petiolulate, usually persistent. Inflorescence axillary, 4-20 cm long, finely puberulous, a few to many-flowered slender thyrse with lower lateral branches up to 8 cm long and widely spreading; pedicels 0.5-1 mm long. Flowers unisexual; calyx patelliform, 0.5-1 mm long, 4-5-lobed, puberulous to subglabrous; petals creamish-white, 3-4(-5), ovate to lanceolate, male 1.7-2 mm long, female 2-2.5 mm long, apex acute, fused for $^1/_2$-$^2/_3$ their length, valvate, erect or slightly spreading in open flower, minutely appressed puberulous outside; staminal tube cyathiform or urceolate, filaments completely fused, male 0.7-1.2 mm long, female ca. 1.7 mm long, margin 7-8-lobed, lobes ca. $^1/_2$-$^3/_4$ length of anthers, glabrous, anthers 7-8, 0.4-0.8 mm long, glabrous; antherodes similar, but without pollen and not dehiscing; nectary absent; ovary ovoid, puberulous, 3-locular, locules with 2 collateral ovules, style-head capitate. Capsule narrowly ellipsoid, 2-3 × 0.7-1 cm, acute at apex and base, apex somewhat rostrate when dry, 3-valved, smooth, finely and densely appressed puberulous; seeds solitary, 1.8-2.3 cm long (including apical arillode), arillode developed along adaxial surface and strongly prolonged at apex, abaxial surface of seed exposed.

Distribution: Southern Suriname, French Guiana and Brazil (Amapá, Pará and Amazonas); lowland non-flooded rain forest; 4 collections studied (SU: 1; FG: 1).

Selected specimens: Suriname: Sipaliwini, vicinity of R. Ulemari, Hammel & Koemar 21404 (K). French Guiana: R. Itany, Toueraki, Moretti 1183 (CAY).

10. **Trichilia pallida** Sw., Prodr. 67. 1788. Type: Hispaniola, Swartz s.n. (holotype S). – Fig. 13

Trichilia echinocarpa (de Vriese) Walp., Ann. Bot. Syst. 2: 227. 1852. – *Portesia echinocarpa* de Vriese, Ned. Kruidk. Arch. 1: 251. 1847. Type: Suriname, Canawapibo, Splitgerber 311 (holotype U [n.v.]).
Trichilia brachystachya Klotsch ex C. DC. in A. DC. & C. DC., Monogr. Phan. 1: 650. 1878. Type. Guyana, Ro. Schomburgk ser. 1, 359 (lectotype BM, designated by Pennington 1981, isolectotypes BR, G, G-DC, GH, K, U, NY, P).
Trichilia davisii Sandwith, Kew Bull. 329. 1933. Type: Guyana, Rupununi R., Davis in Forest Dept. 2167 (holotype K, isotype FHO).

Tree, to 25 m tall (frequently flowering as a small treelet), dioecious; bark of smaller specimens smooth or dippled, pale grey. Young branches pubescent at first, soon becoming glabrous, nearly always with pustular pale small lenticels. Leaves imparipinnate or 3-foliolate, 3-7-foliolate; petiole 2-8 cm long, rachis 2.7-9 cm long, petiole and rachis terete to semiterete, usually glabrous less frequently puberulous to stiffly pubescent; leaflets opposite; petiolule 1-6 mm long; leaflet blade chartaceous to subcoriaceous, usually narrowly or broadly elliptic or oblanceolate less frequently oblong or lanceolate, 6-14 × 3-6 cm, basal leaflets usually smaller, terminal leaflet larger than laterals, apex usually attenuate, acuminate or cuspidate less frequently obtuse, base usually acute or attenuate, puberulous or pubescent on abaxial surface or glabrous, with granular red papillae, sparsely and irregularly glandular-striate or not; venation eucamptodromous or brochidodromous, midrib flat or slightly prominent, secondaries 8-12 pairs, arcuate ascending, parallel, intersecondaries short to moderate, tertiaries often oblique and parallel. Inflorescence axillary or in axils of fallen leaves, 1-3 cm long, fasciculate. Flowers greenish-cream or white, with a yellow nectary, unisexual; calyx usually patelliform or cyathiform, 0.5-1 mm long, lobes 4(-5), ovate or triangular often very shallow, 0.25-1 mm long, apex acute to obtuse, sparsely puberulous to densely pubescent; petals (3-)4, imbricate, elliptic, oblong or lanceolate, 3-5.5 × 1.5-2.5 mm, apex usually acute, usually appressed puberulent outside, inside glabrous or sometimes papillose; staminal tube shortly cylindrical, cyathiform or urceolate, 2-3.5 × 1.5-3 mm, filaments fused for $1/3$-$3/4$ their length, terminated by two acute appendages, outside usually glabrous below and sparsely to densely long hairy above, inside usually glabrous in lower half and sparsely pubescent to densely barbate in throat, anthers (7-)8, 0.6-1 mm long, usually hairy; antherodes narrower, not dehiscing, without pollen; nectary annular or patelliform surrounding base of ovary and fused to base of staminal tube, sparsely pubescent to densely strigose; short androgynophore often present; ovary densely stiff pubescent, (2-)3-locular, locules with 2 obliquely superposed ovules, style slender, glabrous, style-head minute; pistillode vestigal, much shorter than staminal tube, containing vestigial non-functional ovules. Capsule greenish-yellow, ovoid to globose, 0.7-1 cm long, smooth to prominently verruculose or muricate, pubescent, usually with a mixture of longer stiffer hairs, 3(-4)-valved, valves wrinkling horizontally on drying, sometimes strongly reflexed, pericarp ca. 1 mm thick; endocarp thin, cartilaginous; usually only 1 seed developing in each fruit, rarely 1-2 per valve, seed black, ovoid, globose or flattened dorso-ventrally, 0.5-0.6 cm long, shining, arillode red, fleshy, attached to seed along a thin line extending from micropyle along raphe to chalaza, growing over apex, sides, and base of seed, well developed around abortive ovules, seed coat hard.

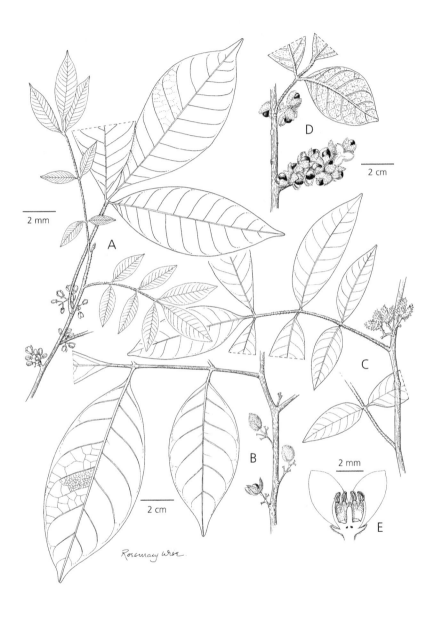

Fig. 13. *Trichilia pallida* Sw.: A, habit with infloresence; B, leaf with smooth fruit; C, leaf with muricate fruit; D, infrutescence; E, ½ flower. (A, E, Evans *et al.* 2433; B, Sabatier & Prévost 2716; C, A.C. Smith 3369; D, Jansen-Jacobs *et al.* 401).

Distribution: From Mexico through C America, the Antilles and tropical S America, the Guianas as far south as northern Argentina and Paraguay; lowland rain forest reaching 800 m alt. in the Guianas; 80 collections studied (GU: 33; SU: 26; FG: 7).

Selected specimens: Guyana: Potaro-Siparuni Reg., Iwokrama Rainforest Reserve, N of Surama Lake within boundaries of Reserve, Ehringhaus *et al.* 55 (US); U. Takutu-U. Essequibo Reg., Rupununi savanna, Natun Bush Island, ca. 2 km SSW of Dadanawa, Gillespie *et al.* 1680 (K); western extremity of Kanuku Mts., in drainage of Takutu R., A.C. Smith 3094 (K, NY, US). Suriname: Kabalebo et Coppename Sinistrum, Florschütz 2819 (U); Sipaliwini, vicinity of Blanche Marie F., on the Nickerie R., Evans *et al.* 2433 (K, US); border Paulus Cr. (lower Suriname R.), Mennega 207 (NY). French Guiana: 14-16 km S of the bridge over Comté R., 59-61 km S of Cayenne, Mori 8889 (K); Saül, La Fumée Mt., on W slope of Antenne Nord, Mori *et al.* 15669 (K); Station des Nouragues, Bassin de L'Arataye, Sabatier 3509 (K, NY), Sabatier & Prévost 2716 (CAY).

Vernacular names: Suriname: kabowasitoko (Arawak), marakatano (Car.), wederhoedoe, witte sali.

Phenology: Flowering and fruiting have been recorded in most months of the year.

11. **Trichilia pleeana** (A. Juss.) C. DC. in Mart., Fl. Bras. 11(1): 215. 1878. – *Moschoxylum pleeanum* A. Juss., Bull. Sci. Nat. Géol. 23: 239. 1830. Type: Brazil, Amazonas, Rio Negro, Plée 10 (holotype P).

Tree, to 20 m, dioecious, bole slightly fluted in larger specimens with small buttresses at base; bark grey, scaling off in long thin irregular sheets which curl up from base. Young branches appressed strigulose at first (trichomes dibrachiate or malpighiaceous) soon becoming glabrous, pale greyish-white or pale brown, lenticellate. Leaves imparipinnate or pinnate with one leaflet of ultimate pair orientated to simulate a terminal leaflet, 5-13(-18) cm long, 3-6-foliolate, with 2-4 much reduced caducous leaflets at base of petiole, these developing before or at same time as normal leaflets and a conspicuous feature of fresh material but often lost on herbarium specimens; petiole and rachis 5-13 cm long, petiole semiterete, rachis semiterete at least in lower part, sometimes canaliculate above, usually glabrous, rarely puberulous; leaflets usually alternate or less frequently opposite; petiolule (1-)3-7 mm long; leaflet blade subcoriaceous, narrowly elliptic or oblanceolate, (7.5-)8-15 × (2.6-)3.2-6 cm, terminal leaflet often slightly larger and basal usually much smaller, apex attenuate or acuminate, rarely acute, base acute or cuneate, glabrous, or occasionally puberulous abaxially on midrib, with scattered appressed (dibrachiate or malpighiaceous) hairs, usually glandular-punctate and -striate, basal reduced leaflets varying from ovate with a strongly oblique base to spathulate or subulate, 0.5-2 cm long, often long petiolulate; venation eucamptodromous, midrib flat or slightly prominent, secondaries 10-14(-16) pairs, ascending, usually slightly convergent or rarely almost parallel, intersecondaries short, obscure or absent, tertiaries obscure. Inflorescence axillary, 9-20 cm long, a few- to many-flowered, narrowly pyramidal thyrse, branches ascending or spreading widely, puberulous to glabrous; pedicels 0.5-1.5 mm long. Flowers unisexual; calyx cyathiform, (0.5-)1-2 mm long, lobes 4-5, aestivation open, irregular, rounded to triangular, acute or margin +/- truncate, sparsely puberulous to glabrous; petals greenish-yellow, 4-5, free or fused $1/4$-$1/2$ their length and then becoming free as corolla lobes reflex, valvate, ovate, lanceolate or narrowly triangular, 3-4.5 × 1-1.5 mm, apex acute, usually papillose or glabrous outside, inside glabrous; staminal tube cyathiform or urceolate, 1.2-2.5 mm long and broad, filaments completely fused, margin with 8-10 subulate lobes alternating with anthers, sparsely appressed puberulous or glabrous outside, crisped puberulous to fairly long pubescent inside especially at throat, rarely glabrous, anthers 8-10, 0.8-1.2 mm long, glabrous; antherodes slender, without pollen; nectary absent; ovary ovoid, sparsely to densely puberulous or appressed puberulous, 3(-4)-locular, locules with 2 collateral ovules, style glabrous, style-head capitate or small discoid at or below level of anthers; pistillode flattened conical containing well-formed, non-functional ovules. Capsule dark green, usually ovoid or globose, less frequently obovoid, 1.5-3 × 1-2.5 cm, usually strongly to weakly verrucose less frequently +/- smooth, subglabrous or glabrous, drying black, (2-)3(-4)-valved, valves remaining +/- erect, pericarp 0.5-3 mm thick, rather tough fleshy, endocarp cartilaginous; seeds 1-2, collateral in each fruit, 1-1.6 × 0.7-1.1 cm, completely surrounded by a thick soft fleshy orange or reddish arillode, otherwise free, seed coat thin and soft.

Distribution: From southern Costa Rica through Panama and northern S America including Trinidad, Venezuela, Guyana and south to central and western Amazonia, Peru and Bolivia, also present in the coastal rain forests of Bahia, Brazil; wet evergreen forest, recorded from both non-flooded and seasonally flooded land, 100-600 m alt.; 9 collections studied (GU: 9).

Selected specimens: Guyana: Rupununi Distr., Kanuku Mts., Crabwood camp, Jansen-Jacobs *et al.* 3471 (US, NY); NW slopes of Kanuku Mts., in drainage of Moku-moku Cr. (Takutu tributary), A.C.

Smith 3549 (K, NY, US); U. Takutu-U. Essequibo Reg., Kwitaro landing, on trail to Shea village, Clarke *et al.* 6218 (CAY, K).

Phenology: Flowering specimens were collected in April, September and October and fruiting specimens from September to January.

12. **Trichilia quadrijuga** Kunth in Humb., Bonpl. & Kunth, Nov. Gen. Sp. 5: 215. 1822. – *Odontandra quadrijuga* (Humb., Bonpl. & Kunth) Triana & Planch., Ann. Sci. Nat., Bot. sér. 5, 15: 374. 1872. Type: Colombia, Magdalena R., Humboldt & Bonpland 1620 (holotype P, isotype P).

In the Guianas only: subsp. **quadrijuga**

Moschoxylum propinquum Miq., Natuurk. Verh. Holl. Maatsch. Wetensch. Haarlem, ser. 2, 7: 74. 1851. – *Trichilia propinqua* (Miq.) C. DC. in A. DC & C. DC., Monogr. Phan. 1: 693. 1878. Type: Suriname, Hostmann 1204a (lectotype U, designated by Pennington 1981, isolectotypes C, F, G, MO, P, S).
Trichilia compacta A.C. Sm., Lloydia 2: 186. 1939. Type: Guyana, Kanuku Mts., A.C. Smith 3545 (holotype NY, isotypes A, B, F, G, K, MO, P, S, US, Y).

Tree, to 30 m tall, dioecious, bole slightly fluted at base in older specimens; bark grey-brown, scaling in long irregular pieces. Young branches with dibrachiate hairs, becoming glabrous, pale brown to greyish-white, rather rough, usually lenticellate. Bud-scales absent. Leaves imparipinnate or pinnate with one leaflet of ultimate pair orientated to simulate a terminal leaflet, (7-)9-11(-13)-foliolate, with a pair of caducous subulate vestigial leaflets of 1-2 mm long and ca. 1-2.5 cm above base of petiole; petiole 2-8 cm long, semiterete, rachis 5-21 cm long, usually canaliculate, minutely puberulous; leaflets alternate to subopposite rarely opposite; petiolule 1-1.5(-3) mm long; leaflet blade chartaceous, oblanceolate, oblong or elliptic, 7-19 × 2-5 cm, apex acuminate or attenuate, base usually rounded, less frequently obtuse, acute, midrib adaxially puberulous to short pubescent, abaxially usually puberulous, abaxial surface with scattered minute dibrachiate hairs or subglabrous, very rarely glandular-punctate and -striate; venation eucamptodromous, midrib sunken, secondaries (11-)14-17 pairs, usually shallowly ascending, arcuate or +/- straight, parallel or slightly convergent, intersecondaries short, obscure, tertiaries obscure. Inflorescence axillary or several clustered on a short axillary shoot, a slender to broadly pyramidal, laxly- to densely-flowered thyrse, 5-20 cm long, flowers +/- erect in bud, strigulose; pedicels 0.5-1.5 mm long. Flowers unisexual, yellowish white; calyx nearly always patelliform rarely rotate or shallowly cyathiform, 0.5-1.5 mm long, lobes (4-)5, broadly triangular or acute, $1/4$-$1/2$ length of calyx, sparsely appressed strigulose; petals 4-5, free or fused for $1/3$-$2/3$ their length and then becoming free as they reflex, valvate, lanceolate, (2-)3-4.5 × 1-1.5 mm, apex acute, sparsely appressed puberulous or strigulose outside, glabrous inside; staminal tube urceolate or cyathiform, 1-2(-2.5) × 1.5-3 mm, filaments completely fused, margin usually with 8-10 lanceolate to subulate lobes alternating with anthers and $1/4$-$3/4$ their length or absent, alternate filaments sometimes shorter, outside usually glabrous, inside glabrous, anthers 8-10, 0.6-0.8(-1.0) mm long, glabrous; antherodes slender, not dehisced, without pollen; nectary absent; ovary ovoid, densely strigulose to short pubescent, 3(-4)-locular, locules with 2 collateral ovules, style stout, glabrous at least in upper half, style-head capitate to small discoid, equalling or exceeding anthers; pistillode +/- conical containing well-formed non-functional ovules. Capsule red, ellipsoid or narrowly obovoid, (1-)2-3 × 0.6-1.3 cm, gradually tapering to base, 3-valved, valves eventually reflexed, apex obtuse to rounded, smooth, densely granular papillose mixed with minute appressed hairs, pericarp 0.3-1 mm thick; endocarp thin, cartilaginous; seeds 1-2, collateral in each fruit, 1-1.5 × 0.5-0.7 cm, broadest above middle, completely surrounded by a thin fleshy, bright red arillode which also develops around aborted ovules, arillode free except for thin line of attachment from micropyle to raphe, seed coat membraneous.

Distribution: Nicaragua to Panama, and widespread across northern S America, the Guianas to coastal Brazil; lowland forest usually on non-flooded land but frequently recorded from riverbanks, 65-900 m alt.; 48 specimens studied (GU: 15; SU: 19; FG: 14).

Selected specimens: Guyana: Potaro-Siparuni Reg., Iwokrama Rain Forest Reserve, Karupukari base camp, Hoffman *et al.* 5045 (US); U. Essequibo Reg., Rewa R., at foot of Spider Mt., Jansen-Jacobs *et al.* 5892 (NY, U, US). Suriname: Nickerie R., Paris Jacob Cr., Maas in LBB 11022 (K); Tumuc Humac Mts., Talouakem, U. Litani R., Acevedo-Rodriguez *et al.* 5857 (K). French Guiana: Approuague R., Arataye R., Barrier *et al.* 2599 (NY); Oyapock R., en aval de Trois Sauts, Oldeman T-976 (US).

Vernacular names: Suriname: meli-sali, soort salie, sorosalie.

Phenology: Flowering mostly from July to November and fruiting from September to April.

Note: The other subsp. *cinerascens* (C. DC.) T.D. Penn. occurs in Nicaragua, Costa Rica and Panama.

13. **Trichilia rubra** C. DC. in Mart., Fl. Bras. 11(1): 203. 1878. Type: Brazil, Pará, Barba, Riedel 1307 (holotype LE, isotypes LE, P, K).

Trichilia guianensis Klotzsch ex C. DC. in A. DC. & C. DC., Monogr. Phan. 1: 657. 1878. Type. Guyana, Ro. Schomburgk ser. II, 794 (= Ri. Schomburgk 1421), (lectotype G-DC, designated by Pennington 1981, isolectotypes G, K, NY, P).

Trichilia guianensis Klotzsch ex C. DC. var. *parvifolia* C. DC., loc. cit. Type: Guyana, Ro. Schomburgk ser. I, 537 (holotype K, isotype BM).

Tree, to 25 m tall (often flowering as a small treelet), dioecious; bark rather rough, mid- to grey-brown, sometimes lenticellate. Young branches appressed puberulous to sericeous, soon becoming glabrous. Leaves imparipinnate, 7-9-foliolate, 9-34 cm long; petiole 2.5-9 cm long, usually narrowly winged, rachis 4-13 cm, semiterete, sparsely puberulous or glabrous; leaflets opposite or subopposite; petiolule 1-3(-4) mm, terminal one sometimes to 2 cm; leaflet blade coriaceous, usually oblong or elliptic, less frequently oblanceolate or lanceolate, 7.5-20 × 2-6 cm, apex acuminate or attenuate, base usually attenuate or acute, less frequently cuneate, often long decurrent into petiolule, especially terminal one, adaxial surface nearly always glabrous, abaxial surface usually glabrous, occasionally sparsely pubescent, sometimes glandular-punctate and -striate; venation eucamptodromous or occasionally brochidodromus, midrib flat or slightly prominent adaxially, secondaries 12-15 on either side of midrib, ascending, usually arcuate and slightly convergent, intersecondaries rather long, secondaries and intersecondaries generally faint. Inflorescence axillary, 1-6 cm long, often several clustered on a short axillary shoot, usually a fascicle or short pyramidal panicle with spreading spikate branches, puberulous; pedicels 0.5-1 mm long. Flowers unisexual; calyx rotate or patelliform, 0.5-1 mm long, lobes 5(-6), broadly ovate or triangular, acute or obtuse, margin ciliate, otherwise glabrous; petals white, cream coloured to yellow, (4-)5, free, imbricate, oblong- or lanceolate, 3-4 × 1-1.5 mm, apex acute to rounded, glabrous; staminal tube cyathiform or urceolate, 1.5-3 × (1-)1.5-2 mm, filaments fused $^1/_3$-$^2/_3$ their length, apex rounded or truncate or less frequently terminated by 2 short acute lobes, glabrous outside, inside glabrous below but barbate at throat, anthers 10, 0.6-0.8 mm long, nearly always densely hairy rarely subglabrous; antherodes slender, not dehiscent, without pollen; nectary absent; ovary ovoid, densely pubescent, (2-)3-locular, locules 1-ovulate, style short, stout, densely pubescent, style-head broadened, discoid; pistillode very slender with smaller non-functional ovules, longer style and a smaller unexpanded capitate style-head. Capsule brownish green, ovoid, ellipsoid or obovoid, 2.5-4 cm long, smooth to occasionally papillose mixed with appressed puberulous indument, 2-(3)-valved, valves sometimes reflexed, pericarp

0.5-1 mm thick, rather soft, endocarp thin-cartilaginous; seeds 1(-2) in each fruit, 1.4-2.2 × 0.8-1.5 cm, completely surrounded by a thin fleshy 3-partite arillode, arillode free except for attachment from micropyle to raphe, seed coat membraneous.

Distribution: A widely distributed but rather sparsely collected species centred in Brazilian Amazonia, extending northwards to Venezuela (delta of the Orinoco R.), Guyana, as far west as Colombia and Peru and southwards to Amazonian Bolivia; almost exclusively recorded from riverbanks on periodically or permanently flooded land, up to 250 m alt.; 23 collections studied (GU: 23).

Selected specimens: Guyana: Mazaruni R., Jenman 802 (K); Waramuri Mission, Moruka R., Pomeroon Distr., de la Cruz 2565 (NY); Pomeroon-Supenaam Reg., Issororo R., 10-12 km W of confl. with Pomeroon R., Hoffman et al. 2703 (K); Essequibo Isl.-W Demerara, lower 7 km of Tiger Cr., Henkel 405 (U); Cuyuni-Mazaruni Reg., Pakaraima Mts., Kurupung R., Takuba Cr. near Kurupung landing, Hoffman et al. 2375 (K).

Vernacular names: Guyana: karababalli, papapee, porokai, sourie, yuriballi (Arawak).

Phenology: Flowering from June to December and fruiting May to September.

Use: Edible fruit (Hoffman 2703).

14. **Trichilia schomburgkii** C. DC. in A. DC. & C. DC., Monogr. Phan. 1: 695. 1878. Type: Guyana, Roraima, Ro. Schomburgk ser. II, 752 (= Ri. Schomburgk 1346) (lectotype G-DC, designated by Pennington 1981, isolectotypes G, K, NY, P).

In the Guianas only: subsp. **schomburgkii**

Trichilia subsessilifolia C. DC. in A. DC. & C. DC., Monogr. Phan. 1: 685. 1878. Type: Suriname, Marowijne R., Kappler 2014 (cited as Hohenacker 2014 by C. DC.) (holotype G, isotypes S, U).
Trichilia cuneifolia Pulle, Recueil Trav. Bot. Néerl. 6: 272. 1909. Type: Suriname, Boschbeheer 78 (holotype U).

Tree, to 20 m tall, dioecious; bark smooth, greyish-brown to greyish-green. Young branches appressed puberulous, becoming glabrous, usually brown, rough, with or without lenticels. Leaves imparipinnate or pinnate with one leaflet of ultimate pair orientated to simulate a terminal leaflet, 5-7-foliolate, 10-25(-35) cm long; petiole semiterete, rachis semiterete below becoming terete above, minutely puberulous to glabrous; leaflets alternate to opposite, heteromorphic, with 2-3 pairs of greatly reduced leaflets clasping base of petiole; petiolule 3-6(-8) mm long; leaflet blade usually coriaceous, oblanceolate less frequently oblong, rarely elliptic, 7.8-27 × 5-8.5 cm, apex attenuate, acuminate, or obtusely cuspidate, base attenuate, both surfaces usually glabrous, glandular-punctate and -striate; reduced basal leaflets unfolding before normal leaflets, orbicular, reniform, 0.4-1 cm long, apex rounded or acute, base often very oblique rounded, petiolulate, usually persistent; venation eucamptodromous to brochidodromous, midrib usually slightly prominent, secondaries (9-)10-14 pairs, ascending, usually slightly arcuate, slightly convergent, intersecondaries short, obscure. Inflorescence axillary, 7-25 cm long, few- to many-flowered thyrse or irregular panicle, puberulous; pedicels 0.5-1.5(-2) mm long. Flowers unisexual, sweetly scented; calyx patelliform, 0.5-1(-1.5) mm long, with 4-5 shallow, broadly ovate or triangular lobes to 1/2 length of calyx or margin +/- truncate, sparsely minute puberulous or glabrous; petals green or greenish cream, 4-5, free, valvate, reflexing, lanceolate or narrowly triangular, 3.5-4.5(-5) × 1-2 mm, apex acute, outside minutely sparsely appressed puberulous, inside glabrous; staminal tube cream, urceolate to short cylindrical, 2-3 × 1.5-2 mm, filaments completely fused, margin with 8(-10) triangular to subulate lobes alternating with anthers and 1/3 to equalling their length, outside usually glabrous rarely sparsely puberulous, inside from sparsely puberulous to coarsely villose, anthers (7-)8(-10), 0.5-1 mm long, glabrous; antherodes slender, not dehiscent, without pollen; obscure nectary at base of ovary (recorded as orange in live specimens); ovary ovoid, 3-locular, locules with 2 collateral ovules, style +/- glabrous, style-head capitate, usually below anthers; pistillode +/- conical with well developed non-functional ovules. Capsule greyish-green, ovoid or ellipsoid, 2-4 × 1.2-2 cm, apex acute or rostrate, smooth, densely appressed puberulous-sericeous, 3-valved, pericarp 0.5-1.5 mm thick; seeds 1-2, collateral in each fruit, 1.3-1.5 × 0.9-1.8 cm, arillode developing around apex and adaxial surface of seed leaving abaxial face exposed, seed coat thin, membraneous. Seedling with spirally arranged 1-foliolate leaves, bearing a single pair of orbicular reduced leaflets at base of petiole.

Distribution: Northern S America including the Guianas; non-flooded lowland forest or periodically flooded forest, from sea level to 800 m alt.; 55 collections studied (GU: 27; SU: 11; FG: 17).

Selected specimens: Guyana: NW Distr., vicinity of Matthew's ridge, Mori 8206 (K); Kamakusa, U. Mazaruni R., de la Cruz 4179 (NY, US); Potaro-Siparuni Reg., Iwokrama International Rain Forest

Reserve, Iwokrama Mts., Clarke 2589 (K, NY, US). Suriname: Tafelberg, Maguire 24195 (K, NY, US); Zanderij I, Stahel 182 (K, NY); Watramiri, Boschwezen 3867 (NY, U). French Guiana: Sinnamary R., Petit Saut - bassin du Sinnamary, Sabatier 2380 (K); Saül, la Fumée Mt., Pennington *et al.* 12142 (K); Cr. Gabaret, 20 km de l'embouchure, 1.5 km sur layon S. Terrain, Oldeman 1925 (K, NY, US); Piste de St. Elie, Acevedo-Rodriguez & Prévost 11093 (CAY).

Vernacular names: Guyana: babonese, karababalli, kulashiré (Arawak), yuriballi. Suriname: hiakanta (Ar.), sorosalie. French Guiana: yooka wiwii, wéti sali.

Phenology: The main flowering season is from July to November and fruiting from October to January, with one record from May.

Notes: *Trichilia schomburgkii* is superficially similar to *T. pleeana* with which it shares the reduced basal leaflets and a similar floral structure; however, it lacks the dibrachiate hairs which occur on the young parts of *T. pleeana* and its fruit and arillode structure is quite distinct. The smooth often rostrate, puberulous or sericeous greyish capsule, containing an apical arillode and exposed seed, contrasts strongly with the dark green verrucose, glabrous capsule of *T. pleeana* which contains a seed completely surrounded by a thin fleshy arillode.

Trichilia schomburgkii subsp. *javariense* T.D. Penn. occurs in western Amazonian Brazil and eastern Colombia.

15. **Trichilia septentrionalis** C. DC. in Mart., Fl. Bras. 11(1): 220. 1878. Type: Brazil, Amazonas, Rio Negro between Manaus and Barcelos, Spruce 1890 (holotype G-DC, isotypes BM, G, K, M, P).

Trichilia moritzii C. DC. in A. DC. & C. DC., Monogr. Phan. 1: 707. 1878. Type: Venezuela, Caracas, Colonia Tovar, Moritz 1681 (lectotype B, designated by Pennington 1981, isolectotypes BM, US).

Tree, up to 20 m tall, monoecious (?); bark smooth, grey to grey-brown, finely longitudinally cracked or lenticellate. Young branches finely puberulous, indument persistent, pale brown, without lenticels. Leaves imparipinnate, 7-11-foliolate; petiole (6.5)15-24 cm long, narrowly winged or semiterete, rachis 6.5-51 cm long, +/- terete, both densely puberulous to glabrous; leaflets opposite; petiolule 3-10 mm long, terminal one often much longer; leaflet blade chartaceous or less frequently subcoriaceous, elliptic, oblong or oblanceolate, (9-)14-26 × (4-)5-13 cm, apex narrowly or broadly short to long acuminate or attenuate, base usually narrowly to broadly cuneate or attenuate rarely +/- truncate, terminal leaflet often with a longer more tapering base than laterals, lower leaflets often much smaller, adaxial surface glabrous or with a few granular red papillae, midrib usually puberoulous, abaxial surface puberulous or glabrous, often intermixed with granular red papillae, often faintly glandular-punctate and -striate; venation eucamptodromous, secondaries 15-22 pairs, ascending, straight or arcuate, usually parallel, intersecondaries short or absent, tertiaries faint to prominent, oblique, +/- parallel. Inflorescence axillary, 10-30 cm long, a laxly-branched panicle, flowers rather densely clustered on lateral branches, usually minutely puberulous; pedicels very stout, to 0.5 mm long or flowers sessile. Flowers greenish cream and sweetly scented, unisexual; male flowers with reduced pistillode and well developed nectary, and female ones with well developed ovary and a reduced or absent nectary; calyx patelliform to cyathiform, 1.5-2.5 mm long, lobes (4-)5, free, strongly imbricate, broadly ovate, orbicular or reniform, apex rounded, puberulous, ciliate outside, glabrous inside; petals 5-6(7), strongly imbricate, usually broadly spathulate or oblong, 4-5.5 × 2-3(-4) mm, apex rounded or obtuse, appressed puberulous to sericeous outside, glabrous inside; staminal tube cyathiform, urceolate or short cylindric, 2-4.5 × 1-2.5 mm, filaments usually completely fused, margin of tube with 6-10 subulate appendages alternating with anthers, outside sparsely hairy in upper half or glabrous, inside sparsely hairy to barbate in throat, anthers 6-10, 1-1.4 mm long, glabrous; antherodes narrow, shrunken, not dehiscent, without pollen; nectary in male flowers a thick glabrous annulus fused with base of staminal tube, in female flowers reduced to a small annulus around base of enlarged ovary or absent; ovary ovoid or conical, pubescent or tomentose, (2-)3 (-4)-locular, locules 1-ovulate, style short, rather stout, pubescent or glabrous, style-head +/- discoid, bearing a conical 3-lobed stigmatic area, pistillode usually greatly reduced and embedded in fleshy nectary containing small non-functional ovules, style thin and usually glabrous. Fruits in large axillary clusters to 22 cm long; capsule red, oblong, obovoid or ellipsoid, 1.3-3.2 × 0.9-1.7 cm, apex acute to rounded, base rounded, smooth to slightly verrucose, densely puberulous, 3-valved, valves opening widely and sometimes reflexing, pericarp 1-2 mm thick, endocarp cartilaginous; seeds 1-2 in each fruit, 1-2.5 cm long, surrounded by a thin 3-partite fleshy red arillode which completely covers seed except for small patch at base, seed coat thin and soft.

Distribution: Costa Rica and northern S America, including the Guianas and all the Andean countries south to Bolivia; non-flooded lowland forest, 150-700 m alt.; 26 collections studied (GU: 1; SU: 8; FG: 17).

Selected specimens: Guyana: Potaro-Siparuni Reg., Mt. Ayanganna, E face, Clarke *et al.* 9755 (U). Suriname: U. Coppename R., Schulz 7814

(NY, U); Sipaliwini, vicinity of Ulemari R., Hammel *et al.* 21416 (K); Brownsberg, BW 6591 (NY). French Guiana: Mt. Tortue, Prévost *et al.* 4462 (K); Saül, Layon Limonade, Raynal-Roques 19980 (CAY, U, US); Ouman fou Langa Soula, Bassin du Haut-Marouini, de Granville *et al.* 9181 (K, U); Saül, Mt. Galbao trail, Mori & Gracie 18651 (CAY).

Phenology: Flowering from May to October and fruiting from January to July.

Note: The species is distinctive on account of the pale green undersurface of the leaflets, the rather prominent parallel secondary venation, and the large erect axillary panicles.

16. **Trichilia surinamensis** (Miq.) C. DC. in A. DC. & C. DC., Monogr. Phan. 1: 679. 1878. – *Moschoxylum surinamense* Miq., Natuurk. Verh. Holl. Maatsch. Wetensch. Haarlem, ser. 2, 7: 73. 1851. Type: Suriname, Hostmann 662 (holotype U, isotypes G, GH, K, M).

Trichilia alternans C. DC. in A. DC. & C. DC., Monogr. Phan. 1: 700. 1878. Type: Suriname, Wullschlägel 1333 (lectotype BR; isolectotype GOET, W).

Tree, 3-15(35) m tall, dioecious; bark smooth, brown, lenticellate becoming greyish-white and cracked when older. Young branches with minute appressed dibrachiate hairs at first, soon glabrous. Leaves imparipinnate or pinnate with one leaflet of ultimate pair orientated to simulate a terminal leaflet, 5-7-foliolate; petiole 4.5-12 cm long, semiterete, rachis 5-15.2 cm long, usually canaliculate on upper surface, usually glabrous rarely appressed strigulose; leaflets alternate or subopposite; petiolule 2-5 mm long; leaflet blade chartaceous, oblanceolate, oblong or elliptic, 10-25 × (3-)4-9 cm, terminal often larger, basal pair usually smaller, apex usually acuminate, base usually attenuate, glabrous or rarely midrib sparsely puberulous adaxially, glandular-punctate and -striate or not; venation eucamptodromous, midrib prominent above, secondaries 11-13 pairs, strongly arcuate ascending, convergent, intersecondaries usually rather short and obscure. Inflorescence axillary, 19-35 cm long, a slender to broadly pyramidal thyrse, branches widely spreading, laxly-flowered, usually subglabrous; pedicels 0.5-1 mm long. Flowers white to greenish-yellow, unisexual; calyx patelliform, 0.5-1 mm long, with 4-5 obtuse or broadly triangular lobes, aestivation open, glabrous; petals 3-4, valvate, remaining erect in open flower, ovate to lanceolate, 2-2.5 × 1-1.5 mm, fused $^1/_2$-$^3/_4$ their length, apex acute, usually glabrous; staminal tube urceolate, 1-1.75 mm long and broad, filaments completely fused, margin with 6-7 subulate lobes alternating with anthers, glabrous, anthers 6-7, 0.6-0.8 mm long, glabrous; antherodes slender, not dehiscent, without pollen; nectary absent; ovary ovoid, minutely appressed puberulous, 3-locular, locules with 2 collateral ovules, style glabrous, style-head minutely capitate; pistillode smaller, containing minute non-functional ovules. Capsule grey-green, ellipsoid to oblong, 1.2-2.2 × 0.7-1.3 cm, smooth, minutely appressed puberulous usually intermixed with granular papillae, 3-valved, pericarp ca. 1 mm thick, endocarp thin, cartilaginous; seeds 1-2, collateral in each fruit, ca. 1.5 cm long, arillode fleshy, strongly developed around apex and along adaxial surface of seed, abaxial face of seed remaining exposed at maturity, seed coat fleshy, thin.

Distribution: Southern Venezuela, the Guianas, to the states of Amapá, Pará and central Amazonia, Brazil; lowland rain forest, often recorded from riparian habitats, usually but not exclusively on non-flooded land, in Suriname up to 1250 m alt.; 79 collections studied (GU: 19; SU: 36; FG: 22).

Selected specimens: Guyana: Kanuku Mts., slope of bank of Guyana, Jansen-Jacobs et al. 1269 (K, NY, US); U. Essequibo Reg., Rewa R., near Corona F., Jansen-Jacobs et al. 5755 (K, NY, US); Rupununi Distr., between Kuyuwini landing and Kassikaityu R., Jansen-Jacobs et al. 2990 (K, U). Suriname: area of Kabalebo Dam project, Nickerie Distr., Lindeman & Görts et al. 510 (U); Juliana Top, 15 km N of Lucie R., Irwin et al. 54752 (K, NY, US); margin of Kayser Airstrip 45 km above confluence with Lucie R., Irwin et al. 57509 (K, NY, US). French Guiana: Reserve des Nouragues, Station de Recherches, Grand Plateau, Poncy 1799 (CAY, NY, US); Saül, Carbet Mais Trail between entrance to Mt. Fumée & Cr. Nouvelle France, Mori et al. 20870 (K, NY); Haut Litani, de Granville 11893 (CAY, K, U); Station des Nouragues, Sabatier & Prévost 2714 (CAY).

Vernacular names: Suriname: melisali, sorosalie, witte sali. French Guiana: akalali (Wayana), sali (Cayenne).

Phenology: Flowering from July to October and fruiting in December.

Note: *Trichilia surinamensis* is closely related to *T. cipo* sharing with it the very small gamopetalous corolla and the seed with a fleshy apical arillode. The best characters for separating *T. cipo* from *T. surinamensis* are the more numerous +/- parallel secondary veins (vs. fewer and strongly arcuate and convergent in *T. surinamensis*), the denser indument of the corolla and calyx (vs. subglabrous in *T. surinamensis*), and their ecology. *Trichilia cipo* frequently occurs along black water streams and in flooded forest, whereas *T. surinamensis* is absent from this habitat.

17. **Trichilia surumuensis** C. DC., Notizbl. Bot. Gart. Berlin-Dahlem 6: 503. 1917. Type: Brazil, Roraima, Serra do Mel, Rio Branco, Ule 8186 (holotype B destroyed, isotypes K, MG).

Tree, up to 15 m tall. Young branches sparsely and minutely strigulose (trichomes basifixed), soon glabrous, bark greyish-white, with a few lenticels. Leaves imparipinnate or pinnate with one leaflet of ultimate pair orientated to simulate a terminal leaflet, 13-22 cm long, 5-6(7)-foliolate; petiole 2-9 cm long, rachis 6-16 cm long, both semiterete, glabrous; leaflets subcoriaceous, usually alternate, rarely opposite, with a pair of minute caducous vestigial scales of 0.5-1 mm long, about 1-1.5 cm above base of petiole; petiolule 3-5 mm long; leaflet blade elliptic or oblanceolate, (8)11-17 × 3.6-7 cm, apex acuminate, base acute to narrowly cuneate, glabrous, glandular-punctate and -striate; venation eucamptodromous to brochidodromous, midrib prominent on adaxial surface, secondaries 13-14 pairs, arcuate ascending, slightly covergent, intersecondaries small, tertiaries obscure. Inflorescence axillary or several on a short axillary shoot, 12-20(-23) cm long, a slender thyrse bearing small clusters of flowers on short lateral branches, glabrous or with a few scattered hairs; pedicels ca. 0.5 mm long. Flowers greenish white with a white staminal tube; calyx shallowly cyathiform, 1-1.5 mm long, with 4 rounded or obtuse lobes, glabrous; petals (3-)4, free, imbricate, spreading, oblong to lanceolate, 3.5-4 × 1.5-2 mm, apex rounded with a few minute appressed hairs on outer surface at tip; staminal tube urceolate or cylindrical, 2-2.5 × 1.5-2 mm, filaments united, margin with 7-8 short acute lobes alternating with anthers and $1/4$-$1/2$ their length, glabrous outside, sparse long hairs in upper half inside, anthers 7-8, ca. 0.8 mm long, glabrous; nectary patelliform, fused to base of staminal tube, glabrous; ovary small, ovoid, puberulous or glabrous, 3-locular, locules with 2 collateral ovules, style long, slender, glabrous, style-head narrowly conical. Capsule dull red, oblong to obovoid, 1.2-2 × 0.6-1 cm, apex rounded, smooth, appressed puberulous, 3-valved, pericarp ca. 0.5 mm thick, endocarp thin-cartilaginous; seeds bright red, solitary, ca. 1.5 × 0.7-0.8 cm, completely surrounded by a fleshy 3-partite arillode, arillode free except at apex and along raphe, seed coat rather thick.

Distribution: Guyana and states of Roraima and Pará, Brazil; forest understorey, up to 715 m alt.; 31 collections studied (GU: 31).

Selected specimens: Guyana: U. Takutu-U. Essequibo Reg., NW Kanuku Mts., along trail through Nappi Cr., Hoffman 3600 (K); U. Takutu-U. Essequibo Reg., 6.5 km NE of Warimure Ranch, SE flank of Kanuku Mts., Harris et al. 1068 (CAY, K); Rupununi Distr., Kanuku Mts., Crabwood Cr., Jansen-Jacobs et al. 3324 (K, U); Upper Essequibo Reg., Rewa R., near Corona F., Jansen-Jacobs et al. 5798 (K, U, NY, US).

Phenology: Flowering in March/April and from August to October, fruiting from December to April.

TAXONOMIC AND NOMENCLATURAL CHANGES

Dilleniaceae

None

Vitaceae

New synonyms:

Cissus erosa Rich. subsp. *linearifolia* (Baker) Lombardi to Cissus erosa Rich.

Cissus boliviana Lombardi to Cissus tinctoria Mart.

Meliaceae

Lectotypification:

Guarea cinnamomea Harms

NUMERICAL LIST OF ACCEPTED TAXA

Dilleniaceae

1. Curatella L.
   1-1 C. americana L.

2. Davilla Vand.
   2-1 D. alata (Vent.) Briq.
   2-2 D. kunthii A. St.-Hil.
   2-3 D. lacunosa Martius
   2-4 D. nitida (Vahl) Kubitzki
   2-5 D. rugosa Poiret   var. rugosa
   2-6 D. steyermarkii Kubitzki

3. Doliocarpus Rolander
   3-1 D. amazonicus Sleumer
   3-2 D. brivipedicellatus Garcke subsp. brevipedicellatus
   3-3 D. dentatus (Aubl.) Standl.
   3-3a subsp. dentatus
   3-3b subsp. esmeraldae (Steyerm.) Kubitzki
   3-4 D. graciilis Kubitzki
   3-5 D. guianensis (Aubl.) Gilg.
   3-6 D. macrocarpus Mart. ex Eichl.
   3-7 D. major Gmel. subsp. major
   3-8 D. multiflorus Standl.
   3-9 D. paraensis Sleumer
   3-10 D. sagotianus Kubitzki
   3-11 D. savannarum Sandw.
   3-12 D. spraguei Cheesm.

4. Neodillenia Aymard
   N. species A

5. Pinzona Mart. & Zucc.
   5-1 P. coriacea Mart. & Zucc.

6. Tetracera L.
   6-1 T. asperula Miq.
   6-2 T. costata Mart. ex Eichl.
        var. costata
        var. rotundifolia (J.E. Smith) Kubitzki
   6-3 T. maguirei Aymard & Boom
   6-4 T. surinamensis Miq.
   6-5 T. tigarea de Candolle

6-6   T. volubilis L. subsp. volubilis
6-7   T. willdenowiana Steud.
6-7a   subsp. emarginata Kubitzki
6-7b   subsp. willdenowiana

Vitaceae

1. Cissus L.
   1-1. C. alata Jacq.
   1-2. C. amapaensis Lombardi
   1-3. C. descoingsii Lombardi
   1-4. C. duarteana Cambess.
   1-5. C. erosa Rich.
   1-6. C. haematantha Miq.
   1-7. C. nobilis Kuhlm.
   1-8. C. spinosa Cambess.
   1-9. C. surinamensis Desc.
   1-10. C. tinctoria Mart.
   1-11. C. trigona Willd. ex Schult. & Schult.f.
   1-12. C. ulmifolia (Baker) Planch.
   1-13. C. venezuelensis Steyerm.
   1-14. C. verticillata (L.) Nicolson & C.E.Jarvis subsp. verticillata

Meliaceae

1. Azadirachta A. Juss.
   1-1. A. indica A. Juss.

2. Cabralea A. Juss.
   2-1. C. canjerana (Vell.) Mart. subsp. canjerana

3. Carapa Aubl.
   3-1. C. akuri Poncy, Forget & Kenfack
   3-2. C. guianensis Aubl.
   3-3. C. surinamensis Miq.

4. Cedrela P. Browne
   4-1. C. fissilis Vell.
   4-2. C. odorata L.

5. Guarea F. Allam. ex. L.
   5-1. G. carinata Ducke
   5-2. G. cinnamomea Harms
   5-3. G. convergens T.D. Penn.
   5-4. G. costata A. Juss.

5-5.   G. glabra Vahl subsp. glabra
5-6.   G. gomma Pulle
5-7.   G. grandifolia DC.
5-8.   G. guidonia (L.) Sleumer
5-9.   G. kunthiana A. Juss.
5-10. G. macrophylla Vahl
5-10a. subsp. pachycarpa (C. DC.) T.D. Penn.
5-10b. subsp. pendulispica (C. DC.) T.D. Penn.
5-11. G. michel-moddei T.D. Penn. & S.A.Mori
5-12. G. pubescens (Rich.) A. Juss.
5-12a. subsp. pubescens
5-12b. subsp. pubiflora (A. Juss.) T.D. Penn.
5-13. G. scabra A. Juss.
5-14. G. silvatica C. DC.
5-15. G. trunciflora C. DC.
5-16. G. velutina A. Juss.

6. Khaya A. Juss.
   6-1.   K. senegalensis (Desr.) A. Juss.

7. Melia L.
   7-1.   M. azedarach L.

8. Swietenia Jacq.
   8-1.   S. macrophylla King

9. Trichilia P. Browne
   9-1.   T. areolata T.D. Penn.
   9-2.   T. cipo (A. Juss.) C. DC.
   9-3.   T. elegans A. Juss. subsp. elegans
   9-4.   T. euneura C. DC.
   9-5.   T. lecointei Ducke
   9-6.   T. lepidota Mart. subsp. leucastera (Sandwith) T.D. Penn.
   9-7.   T. martiana C. DC.
   9-8.   T. micrantha Benth.
   9-9.   T. micropetala T.D. Penn.
   9-10. T. pallida Sw.
   9-11. T. pleeana (A. Juss.) C. DC.
   9-12. T. quadrijuga Kunth subsp. quadrijuga
   9-13. T. rubra C. DC.
   9-14. T. schomburgkii C. DC. subsp. schomburgkii
   9-15. T. septentrionalis C. DC.
   9-16. T. surinamensis (Miq.) C. DC.
   9-17. T. surumuensis C. DC.

COLLECTIONS STUDIED

(Numbers in **bold** represent types)

Dilleniaceae

GUYANA

Abraham, A.A., 21 (2-2); 70 (1-1)
Anderson, C.W., 1 (2-2); 438 (2-1);
    507 (6-1)
Archer, W.A., 2325 (2-2)
Atkinson, D.J., 110 (5-1)
Austin, D.F., 7745 (3-7)
Bartlett, A.W. , s.n. (6-1)
Beckett. J.E. & P. Kortright, 8507(2-2)
Boom, B.M. *et al.*, 7118 (6-1); 7187
    (3-3a); 7273 (2-4); 7351 (2-4);
    7546 (3-11); 9308 (3-12);
Clarke, H.D. 4837 (3-9)
Cook, C.D.K. 29 (1-1)
Cowan, R.S. , 1802 (6-1); 1806 (3-12);
    2045 (2-2); 2111 (3-5); 39244 (6-1)
Cruz, J.S. de la, 1619 (2-2); 1779 (6-
    1); 2230 (6-1); 2371 (6-4); 2644
    (6-1); 2667 (6-1); 2673 (3-2); 2866
    (6-1); 2890 (6-4); 3069 (6-1); 3469
    (2-2); 3593 (2-2); 3595 (2-2)
Dans, D.M. 181 (3-12)
Davis, D.H., 56 (6-1); 631 (2-2); 1977
    (2-2); 1981 (6-1)
Ek, R.C. *et al.*, 742 (3-2a); 836 (3-5);
    838 (3-12); 876 (3-12);
Fanshawe, D.B., F-1179=3915 (6-2);
    F-1790 (2-1)
Forestry Department Guyana 2404=D-
    408 (6-1); 2651= F-52 (6-2);
    2953 (1-1); 3915 =F-1179 (6-2a);
    5598=F-2799 (2-2); 5961=WB-565
    (2-2); 6000=F-2853 (5-1) 6283 (1-
    1); 7021=F-3443 (5-1);
Gelson, M., 157 (6-1)
Gillespie, L.J . *et al.*, 996 (3-3a); 1266
    (2-2); 1504 (3-2a); 1830 (2-4);
    2153 (2-2); 2552 (2-2); 2672 (3-11)

Gleason, H.A., 201 (2-2); 669 (2-2);
    746 (6-1); 756 (6-1); 813 (6-1);
Goodland, R.J.A., 745 (1-1); 950 (2-
    4); 968 (2-2)
Görts-van Rijn, A.R.A., 393 (2-5); 398
    (2-2)
Graham, E.H. 408 (2-2)
Graham, V. 540 (2-2)
Hahn, W. *et al.*, 3933 (3-12); 4438 (3-
    12); 5662 (2-2)
Hancock, J. , 293 (6-1)
Harris, S.A., A-7 (6-1)
Harrison, S.G., 742 (6-1); 576 (2-2);
    759 (2-2)
Harrison, S.G. & R. Persaud, 1053 (6-1)
Henkel, T.W. *et al.*, 1108 (3-11); 1835
    (3-5); 2030 (3-5); 2064 (3-12);
    2564 (2-2); 3702 (2-2); 4942 (5-1);
    5851 (3-5)
Hill, S.R., 27217 (6-1); 27117 (6-1);
    27280 (1-1); 27198 (6-1)
Hitchcock, A.S., 17307 (6-1)
Hoffman, B. *et al.*, 880 (6-7b); 964 (2-
    5); 1182 (3-5); 3382 (3-11); 3721
    (2-4), 3879 (1-1);
Hohenkerk, L.S. 656 (6-1); 656A (6-1)
Hutchinson, I., 350Q2N2 (6-7)
Irwin Jr., H.S. BG-9 (6-1); 130 (3-12);
    201 (2-2); 305 (1-1); 359 (3-12)
Jansen-Jacobs, M.J. *et al.*, 228 (3-5);
    1031(2-2); 1797 (2-1); 2052 (1-1);
    2293 (3-9); 2493 (5-1); 2869 (3-
    12); 3235 (2-2); 3622 (2-2); 5585
    (3-12)
Jenman, G.S. , 303 (6-1); 850 (2-2);
    1038 (3-11); 1233 (6-1); 2146 (2-2);
    3818 (6-1); 4687 (2-2); 4746 (2-2);
    4897 (6-4); 4938 (6-1); 5580 (1-1);
    6225 (2-2); 6655 (2-2); 6676 (2-2)

Kelloff, C.L. *et al.*, 925 (3-12); 1159 (1-1)

Kress, W.J., 86-1871 (1-1)

Lee, G. & R. Persaud , 20 (6-1)

Maas, P.J.M *et al.* 3514 (6-1); 3645 (1-1); 4043 (1-1); 4381 (3-11); 5472 (3-7a); 5475 (2-2); 5540 (3-7a); 5846 (3-11)

Maguire, B. 23013 (6-1); 23066 (5-1); 23152 (3-11); 23207 (2-2); 23318 (3-12); 23417 (3-12); 32627 (3-11); 32651 (3-11); 35047 (2-2); 40620 (6-3); 40638 (6-3); 40650 (3-11); 40670 (6-3); 43807 (3-11)

Martyn, E.B., 136A (6-1); 138 (2-2); 393 (2-2)

McDowell, T. *et al.*, 1840 (1-1); 2049 (3-2a); 2079 (3-5); 2341 (3-5); 2461 (3-3a); 2506 (1-1); 2546 (3-12); 2775 (2-2); 2790 (3-12); 3537 (2-2); 4120 (2-2); 4491 (2-2)

Mori, S. *et al.*, 8042 (3-12)

Parker, C.S., 299 (2-2)

Pennington, R.T., 391 (6-1)

Persaud, A.C., 15 (2-2); 28 (3-6)

Pipoly, J.J. *et al.*, 7598 (2-2); 7989 (6-1); 9144 (6-1); 9577 (6-1); 9406 (6-6a); 9480 (6-4); 9569 (6-1); 9239 (6-1); 11355 (2-2); 11384 (2-2); 11387 (6-1); 11481A (1-1); 11484 (3-12); 11487 (1-1); 11395 (3-12); 11515 (1-1); 11502 (2-2); 11558 (2-2); 11565 (3-12)

Robertson, K.R., 262 (2-2)

Sandwith, N.Y., 479 (6-4); 686 (3-7a); 1053 (2-2); 1377 (3-11);

Skog, L.E. & C. Feuillet, 7117 (2-5)

Smith, A.C., 2184 (2-2); 2185 (1-1); 2756 (3-9); 2846 (3-2a); 2903 (5-1); 3101 (2-2); 3273 (3-3a)

Steege, H.S. ter, 395 (2-2)

Stockdale, F.A., s.n. (2-5); s.n. (2-2); 8791 (6-2b)

Stoffers, A.L., 403 (2-2); 453 (2-2); 493 (3-3a)

Tate, G.H.H. 57 (1-1)

Tutin, T.G. , 251 (3-12)

U.G. Bio: 106, s.n. (3-2a); 13 (6-1); 72 (6-1)

U.G.Field Group (1970), 12 (6-1)

Williams. R.D., 2 (2-2)

SURINAME

Andel, T.R. van *et al.*, 5408 (3-4)

BW (Boschwezen Suriname) 11 (2-1); 1217 (1-1); 2069 (2-2); 4731 (1-1)

Boerboom, J.H.A., 9198 (2-2); 9188 (2-5); LBB 11481 (2-2)

Breteler, F.J., 3645 (1-1)

Budelman, A. 795 (2-2); 1389 (2-5); 1391 (2-5);

Bureau, F., 5485 (6-1); 6554 (2-4); 6822 (2-2)

Donselaar, J. van *et al.*, 93 (6-1), 285 (6-1); 1666 (2-1); 2045 (3-8); 3219 (2-1); 3864 (3-6)

Evans, R. *et al.* 1812 (3-12); 1854 (2-2); 1920 (2-2); 1948 (3-9); 2357 (6-5); 2608 (6-1); 2825 (3-9); 2896 (3-4); 3210 (3-5); 3226 (3-12);

Everaarts, A.P., 606 (6-1)

Florschütz, J. & P.A. 830 (6-1)

Focke, H.C. 38708 (6-4)

Geyskes, D.C., 160 (2-2)

Hammel, B. *et al.*, 21629 (3-4); 21273 (3-5)

Hawkins, T. 2151 (3-12);

Heyde, N.M., 399 (3-7a); 552 (3-2a); 584a (3-12); 702 (2-2); 722 (3-2a)

Heyligers, P.C., 556 (6-1)

Hostmann, F.W., 1005 (6-1); 1141 (6-1); 38709 (6-2b)

Irwin, H.S., 55221 (3-12); 57532 (2-2)

Jansma, R. , 15612 (2-2)

Jongking, C., UVS-16872 (2-5); UVS-16873 (2-2); UVS-17062 (2-1)

Kappler, A., 1711 (2-5)

Koppert, T., 69 (3-12);

Kramer, K.U. & W.H.A. Hekking, 2466 (6-1); 2838 (3-9)

Krukoff, B.A., 12298 (3-2a)

Lanjouw, J. & J.C. Lindeman, 576 (2-2); 842 (6-1); 1782 (6-1); 1910 (3-7a); 2079 (3-7a); 2200 (3-9); 2665 (3-12); 2821 (3-4); 2987 (6-1); 3258 (3-12); 3260 (3-12);

Lems, K., 640218 (2-2)

Lindeman, J.C. et al., 45 (2-2); 52 (2-1); 232 (3-3a); 265 (2-2); 415 (6-1); 1956 (3-2a); 4131 (3-3b); 5198 (5-1); 5301 (2-2)

LO – Studenten, 15754 (2-2)

Lohmann, L.G. et al., 232 (3-1a)

Maas, P.J. M. et al., 3243 (1-1); 3297 (2-2); 10819 (3-5)

Maguire, B., 23630 (2-2); 23653 (3-12); 23694 (6-1); 23763 (2-2); 23782 (3-7a); 23782b (3-7); 23794 (3-7a); 23863 (3-7a); 24284 (3-12); 24285 (6-1); 24396b (6-1); 24638 (6-1); 24704 (6-1); 24965 (6-1); 25009 (3-7a); 40832 (2-2); 40834 (2-5)

Mennega, A.M.W., 244 (3-7a)

Oldenburger, F.H.F. et al., ON-295 (2-2); ON-300 (1-1); ON-456 (2-5)

Pulle, A.A., 524 (5-1)

Rozewijn, A. & O. Paulus, C-10744 (2-2)

Ruysscharert, S. et al., 192 (3-9); 790 (3-6); 808 (3-2a); 950 (3-6)

Samuels, J.A., s.n. (6-1); s.n. (3-7a); 144 (2-2); 532 (3-12);

Sauvain, M., 702 (2-5); MS-574 (3-5)

Schulz, J.P., 8529 (2-2); 8642 (2-2)

Wessels Boer, J.G., 267 (3-7a); 442 (3-4); 1362 (3-12);

Stahel, G., 209 (1-1)

Vreden, C. LBB-11285 (2-5)

Wigman, J.R., 50 (3-7a)

FRENCH GUIANA

Acevedo, P. 4944 (2-1)

Allorge, L. 211 (3-4)

Aublet, J.B.C.F., s.n. (3-12);

Aubreville, A. 262 (2-2)

Barbier, M. s.n. (6-5)

Benoist, R. s.n. (6-1); 488 (6-2a); 736 (2-1); 842 (5-1)

Billiet, F. 1041 (2-3); 1455 (1-1)

Billiet, F. & B. Jadin, 4494 (3-4)

Bordenave, B.G. 797 (3-5); 833 (6-6a); 821 (2-1)

Broadway, W.E. 540 (6-5); 483 (1-1); 753 (2-2); 504 (6-5)

Brown, R. s.n. (6-2b)

Capus, F. 158 (2-3)

Chaix, M. 13 (2-2)

Cowan, R.S. 38035 (6-1)

Cremers, G.A. et. al. 4897 (2-3); 7325 (2-3); 9426 (3-12); 9555 (2-2); 9628 (6-1); 11254 (2-5); 12458 (6-1); 12459 (2-2)

Descoings, B. 20196 (2-3); 20254(2-3)

Feuillet, C. 606 (2-2); 606 p.p. (2-3); 614 (3-7a)3571 (2-2); 3016 (6-1); 4610 (2-1); 1467 (2-2)

Fleury, M. 614 (3-7a); 1430 (3-7a)

Forest Service 7016 (2-5); 7033 (2-5)

Foresta, H. de 626 (3-7)

Garnier, 187 (2-3)

Granville, J.J. de et al., 1803 (3-2); 4865 (2-2); 8886 (3-5); 9112 (3-5); 9628 (3-7a); 10789 (5-1); 10948 (3-5); 15630 (2-2);

B-4250 (2-5); B-5319 (2-2)

Grenand 1829 (2-3)

Hoff, M. 5665 (3-7a); 5356(2-4)

Jacquemin, H. 1792 (2-2); 1985 (2-1)

Jussieu, d'Adrien de 280(6-5)

Kodjoed, J.F. 134 (2-1)

Leblond, M. 379 (6-5)

Lemee, M.A. s.n. (6-1); s.n. (2-5); s.n. (2-2); s.n. (6-5)

Lemoine, M. 7826 (3-12); 7905 (3-7a)
Leprieur, F.M.R. s.n. (2-2); s.n. (1-1);
s.n. (6-5); s.n. (6-1)
Lescure, J.P. 147 (3-5); 261 (2-2)
Loubry, D. , 1663 (3-4); 1758 (2-1)
Martin s.n. (6-5)
Maas, P.J.M. *et al.*, 2212 (3-2a)
Mélinon, E.M., s.n. (1-1); s.n. (2-2);
s.n. (6-2a); s.n. (2-1); 66 (5-1); 85
(2-5); 90 (6-1); 1864 (2-2); 400 (2-
2)
Mori, S. *et al.*, 8866 (2-1); 15158 (2-
2); 23428 (3-3b); 23434 (3-3b);
23515 (3-3b); 23583 (3-12); 23689
(3-5); 24153 (3-2a)
Oldeman, R.A.A., B-1255 (2-2);
B-2403 (2-2); B-2683 (2-1); B-2849
(3-10); B-2909 (2-2); B-3231 (3-
5); B-3427 (3-9); B-3920 (2-5);
B-3930 (2-3); B-3936 (1-1); 1280
(2-2); 1408 (3-2a); 1454 (2-2);
1978 (2-5); 2403 (6-5);
Perrottet, G.S. sn (2-2)
Perrottet, M. sn (2-2); 1820 (2-2)
Petitbon 152 (2-2)
Prévost, M.F. *et al.*, 496 (2-5); 1399
(3-7a); 1791 (2-2); 3093 (3-5);
3551 (2-2); 3921 (2-2);
Raynal-Roques, A. 19900 (6-1)
Richard, Louis Claude s.n. (6-1); s.n.
(6-2a); s.n. (5-1)
Riero 1027 (2-2)
Sabatier, D. *et al.*, 813 (2-1); 928 (3-
5); 994 (3-7a); 5350 (5-1)
Sagot, P.A., s.n. (6-2a); 3 (6-4); 1155
(3-10)
Sagot, R. s.n. (6-5); 14 (6-1); 15 (6-4);
17 (2-5); 1261 (2-2); 1219 (6-5);
5741 (6-2a)
Sastre, C. *et al.* 216 (6-1); 249 (2-3);
3909 (3-9); 4005 (3-7a); 4681 (4-
1); 6319 (3-2);
Sauvain, M. 633 (5-1)
Schnell, R. 11241 (2-5); 11270 (2-2)

Service Forestrier 7972 (2-5); 6224 (2-
3); 4028   (2-2)
Talbot, H.F. s.n. (2-1)
Toriola-Marbot, D. & M. Hoff , 49 (3-
7a); 119 (1-1)
unk. 237 (2-2); 1889 (6-4)
Veyret, Y. 4399 (2-3)
Weitzman, A. & W. Hahn 279 (2-2)

Vitaceae

GUYANA

Andel, T.R. van, *et al.*, 672 (1-14);
1133 (1-3); 1312 (1-14); 1657 (1-
3); 1803 (1-14).
Archer, W.A., 2359 (1-5).
Clarke, H.D., 283 (1-14); 300 (1-
5); 465 (1-14); 1975, 2111, 2220,
3526, 3825 (1-5); 4634 (1-3); 4830,
5974 (1-5).
Cramer, J.M., 12 (1-5).
Cruz, J.S. de la, 1409 (1-14); 1759,
2030, 2346 (1-5); 2375, 3010 (1-
14); 3098 (1-5); 3156 (1-14); 3225
(1-5); 3263 (1-14); 3743, 3974 (1-
5); 4090, 4446 (1-14).
Davis, D.H., 30 (1-5).
Ek, R.C., *et al.*, 728 (1-3).
Fanshawe, D.B., 2051 (1-6); 3361 (1-
13)
FD (Forest Dept. British Guiana),
5118 (1-14).
Gillespie, L.J., 760 (1-5); 2181 (1-3).
Görts-van Rijn, A.R.A., *et al.*, 289 (1-
5).
Grewal, M.S., *et al.*, 207 (1-5); 266 (1-
8); 351 (1-5).

Hahn, W.J., *et al.*, 3812 (1-8).
Henkel, T.W., *et al.*, 618 (1-5); 2634
(1-14); 3012 (1-5); 3074 (1-14);
5315 (1-5); 5316 (1-6).
Hitchcock, A.S., 16531, 16648 (1-14);
16878 (1-5).

Hoffman, B., *et al.*, 1343 (1-5); 2494, 2716 (1-14).

Im Thurn, E.F., s.n. (1-14).

Jansen-Jacobs, M.J., *et al.*, 730 (1-14); 1557 (1-5); 2593, 2806, 3173, 3694, 3828, 3899 (1-5); 4216 (1-14); 4490 (1-1); 4796 (1-4); 5681 (1-14); 5799 (1-6).

Jenman, G.S., 149, 461 (1-14); 1503 (1-5); 4827, 5351 (1-14); 5352 (1-5).

Kelloff, C., *et al.*, 643 (1-5).

Kvist, L.P., *et al.*, 376 (1-5); 403 (1-14).

Lall, H., 311 (1-8).

Maas, P.J.M., *et al.*, 3518, 5557 (1-14); 4085, 7171 (1-5); 7258 (1-4).

Maguire, B., *et al.*, 22991 (1-14).

McDowell, T., *et al.*, 1776, 1858 (1-5); 2356, 2419 (1-14); 3285 (1-5).

Mutchnick, P., 534 (1-14); 624, 1541 (1-5).

[6] Parker, C.S., s.n. (1-8).

Persaud, C.A., 231 (1-8); 274 (1-14); 275 (1-5); 335 (1-14).

Pipoly, J.J., 8224, 8280, 8380, 9075, 9247 (1-14).

Sandwith, N.Y., 1054 (1-5).

Smith, A.C., 3506 (1-14); 3520 (1-5).

Steege, H. ter, 222 (1-5).

Tillett, S.S., *et al.*, 45890 (1-13).

SURINAME

Angremond, A. d', s.n. (1-5); s.n. (1-14).

Berthoud-Coulon, M., 528 (1-14); 568 (1-5); 569 (1-8); 570, 571 (1-14).

Boon, H.A., 1070, 1255 (1-14).

Bordenave, B.G., *et al.*, 8080 (1-14).

Borsboom, N.W.J., 11994 (1-5).

BBS (Bosbeheer Suriname), 233 (1-5).

BW (Boschwezen) , 751 (1-5); 1019 (1-14); 2143 (1-5); 2853 (1-9); 3516 (1-14); 5272 (1-5); 5274 (1-14); 5317 (1-5); 5611 (1-14).

[7] Daniëls, A.G.H. & F.P. Jonker, 1323 (1-5).

Donselaar, J. van, 1347 (1-14).

Essed, E., 3b (1-14).

Evans, R., *et al.*, 3398 (1-5).

Everaarts, A.P., 552, 1085, 1123, 1147 (1-14).

Florschütz, P.A., *et al.*, 150, 894 (1-5); 1119 (1-14); 1404 (1-5); 1607 (1-11); 1929 (1-14); 2248 (1-5); 2249 (1-14); 2405, 2460, 2463 (1-14).

Focke, H.C., 688 (1-14).

Granville, J.J. de, *et al.*, B-4543, 11944 (1-5).

Hammel, B. & S. Koemar, 21246 (1-5); 21260 (1-14).

Hekking, W.H.A., 824 (1-14).

Herb. Linnaeus, 149.5 (1-5).

Heyde, N.M., *et al.*, 108 (1-6); 227, 325 420, 509 (1-14).

Hostmann, F.W.R., 129 (1-5); 210 (1-14); **1301** (1-5).

Hugh-Jones, D.W., 54 (1-14).

Hulk, J.F., 288 (1-5).

Irwin, H.S., *et al.*, 55313 (1-5); 55403 (1-14); 55464, 55838, 55924, 57584 (1-5).

Jansen-Jacobs, M.J., *et al.*, 6174 (1-14); 6175 (1-5).

Jongh, E.G., 30 (1-5).

Jonker-Verhoef, A.M.E. & F.P. Jonker, 46, 176 (1-8); 433, 665 (1-14).

Kappler, A,. 129, 1598 (1-5); **1959**, s.n. (1-6); s.n. (1-5).

---

[6] Cited elsewhere as M. or W. Parker, erroneously

[7] Collection number following Jonker-Verhoef's numbering, but erroneously labeled as Daniels & Jonker

Kock, C., s.n. (1-5).
Kramer, K.U. & W.H.A. Hekking, 2027 (1-5); 2619 (1-14); 2710 (1-5); 2711 (1-14); 2976 (1-5); 3134 (1-14).
Kuyper, J., 532 (1-14).
LBB (Lands Bosbeheer), 10537 (1-5); 10538, 10650 (1-14); 12397, 12428, 12614, 12703 (1-14); 12716 (1-5); 13203 (1-5); 13219 (1-14); 13771, 14154 (1-8).
Lanjouw, J., 271 (1-8); 529 (1-5); 592 (1-8); 819 (1-5); 716, 1104 (1-14); 1222 (1-5).
Lanjouw, J. & J.C. Lindeman, 1116 (1-8); 1386, 1458 (1-14); 1485 (1-8); 1506, 2065, 3099 (1-14); 3419 (1-8); 3467 (1-14).
Lindeman, J.C., 4425, LBB 12115 (1-5).
Lindeman, J.C. & A.R.A. Görts-van Rijn, et al., 101, 173, 351 (1-14); **457** (1-9).
Lindeman, J.C. et al; 138, 225, 664 (1-14).
Maguire, B., et al., 22708 (1-14); 22785, 23769 (1-5); 23787 (1-14); 53947, 54233 (1-5).
Mennega, A.M.W., 77 (1-5); 193 (1-14); 212 (1-5); 243, 387 (1-14).
Miller, J.S. & W.D. Hauk, 9248 (1-14).
Narain, T.R., LBB 13771, LBB 14154 (1-8).
Oldenburger, F.H.F., et al., 462 (1-14); 697 (1-5); 809 (1-4); 822 (1-5); 831 (1-4).
Palmer-Jones, R.W., 17 (1-5).
Plotkin, M., Z-05606 (1-7).
Poncy, O., et al., 1186 (1-14).
Pons, T.L, LBB 12703 (1-14); LBB 12716 (1-5).
Pulle, A.A., 23 (1-5); 94, 178 (1-14); 224, 255 (1-5); 288 (1-14).
Reijenga, T.W., 517, 518, 599 (1-14).

Rombouts, H.E., 54 (1-14); 441 (1-4); 501 (1-5); 504 (1-4); 651 (1-14); 671, 727 (1-5).
Rosário, C.S., et al., 1756 (1-5); 1885 (1-10); 1927 (1-14); 2201 (1-10).
Samuels, J.A., 255, 314, 462 (1-5); 504 (1-14).
Sastre, C., 1395 (1-5).
Sauvain, M., 595 (1-14).
Schimper, A.F.W., 129 (1-5).
Schulz, J.P. et al., LBB 10537 (1-5); LBB 10538, LBB 10650 (1-14).
Soeprato, 12G (1-14); 35E (1-5); 110, 130, 278 (1-14); 302 (1-5).
Splitgerber, F.L., 233 (1-5); 1142 (1-14); 1143, s.n. (1-5).
Stahel, G., 349 (1-5); 615 (= BW 7170) (1-14).
Sterringa, J.T., LBB 12397, LBB 12428 (1-14).
Teunissen, P.A, et al., LBB 12614 (1-14); LBB 13203 (1-5); LBB 13219 (1-14).
Tresling, J.H.A.T., 125 (1-14); 267 (1-5).
Tuinzing, J., 1964 (1-14).
Tulleken, J.E., 120 (1-5); 159, 301 (1-14); 420 (1-5); 424 (1-14).
Versteeg, G.M., 58 (1-14); 59 (1-5); 676, 895 (1-14).
Wessels Boer, J.G., 228 (1-5); 231, 534, 556 (1-14); 742 (1-5); 757 (1-4).
Wezelt, s.n. (1-14).
Wullschlägel, H.R., 64 (1-5); 65 (1-14); **66** (1-8).

FRENCH GUIANA

Arquembourg, S. & J. Dervaux, 30 (1-5).
Barrabé, L. et al., 36 (1-14); 158 (1-5).
Barrier, S., 2753 (1-9).
Belbenoit, P., 512 (1-14).
Benoist, R., 16 (1-14).

Berton, M.-E., 257 (1-5).
Billiet, F. & B. Jadin, 999, 1268 (1-14); 4668 (1-5); 4741 (1-14).
Bordenave, B.G. *et al.*, 108 (1-14); 144 (1-8); 186 (1-5); 8026 (1-6); 8466 (1-14).
Bourdy, G., 2949 (1-5).
Cadamuro, L., *et al.*, 469 (1-4); 488, 794 (1-14).
Cowan, R.S., 38883 (1-8).
Cremers, G., *et al.*, 4508, 7756 (1-14); 7840 (1-6); 8672 (1-14); 9760 (1-3); 9823, 12193 (1-5); 12440, 13872 (1-6); 14445, 14581 (1-4); 14683 (1-6).
Croat, T.B., 53798 (1-14).
Crozier, F., 114 (1-14).
Delavault, J.J., s.n. (1-14).
Delnatte, C., *et al.*, 355 (1-14); 525, 1111 (1-8).
Feuillet, C., *et al.*, 843 (1-5); 1729 (1-14); 2374 (1-6); 2403 (1-5); 2878 (1-14); **2925**, 3975 (1-3).
Fleury, M., 165 (1-5); 257, 448, 1437 (1-14); 1964 (1-5).
Foresta, H. de, 605 (1-14).
Garnier, F.A., 9 (1-3).
Gonzalez, S. & M. Chaiz, 1518 (1-5).
Granville, J.J. de, *et al.*, 1361 (1-5); 1926, 5755 (1-14); 6858 (1-8); 6921 (1-5); 7007 (1-14); 7277 (1-3); 8210 (1-5); 8306 (1-14); 9160 (1-8); 9290, 9304, 9505, 9921 (1-5); 9960, 10028 (1-14); 11728 (1-8); 12507 (1-5); 12508 (1-14); 13426 (1-4); 14740 (1-8); 14993, 15354, 15904 (1-5); 15586 (1-8); 16116 (1-14); 16281 (1-5); 16369 (1-11); 16441 (1-14); 16837 (1-6); B-4630 (1-5); C-33 (1-5).
Grenand, P., 10, 1623 (1-5).
Hahn, W.J., 3523 (1-14).
Hallé, F., 552 (1-5); 655, 1045 (1-14).
Haxaire, C., 633 (1-5); 649 (1-14).

Hequet, V., 86 (1-4); 296 (1-3).
Herb. A.J.A. Bonpland, s.n. (1-5).
Herb. Richard, s.n. (1-14).
Hoff, M., *et al.*, 5008 (1-14); 5405 (1-5); 5607 (1-4); 6277 (1-6); 6326 (1-3); 8014, 8082 (1-5); 8131 (1-14).
Jacquemin, H., 1513 (1-5); 1583 (1-14); 1584 (1-5); 1605 (1-6); 1724, 1733 (1-5).
Le Goff, A., *et al.*, 32 (1-14); 127 (1-8); 225 (1-5).
Leblond, J.B., **78** (1-5); **79**, s.n. (1-14).
Leclerc, A., 23-bis (1-4).
Lemée, A., s.n. (1-14).
Lescure, J.P., 572 (1-5).
Loubry, D., 1160 (1-5).
Marshall, N. & J. Rombold, 201 (1-5).
Martin, J., **s.n.** (1-5).
Mori, S.A., *et al.*, 18392 (1-5); 23066 (1-14); 23920 (1-11); 24199 (1-3).
[8] Oldeman, R.A.A., 1074 (1-14); 1312 (1-5); 1742, 2318, 2663 (1-14); B-494, B-805 (1-5); B-957, B-1425 (1-5); B-1429 (1-6); **B-2501** (1-2); B-2821, B-2866 (1-14); B-3057 (1-3); B-3214 (1-14); B-3259 (1-5); B-3309 (1-14); B-3610 (1-5); BC-2 (1-5); BC-30 (1-14); T-183, T-328, T-657 (1-5); T-774 (1-14).
Pérez, A., *et al.* 945 (1-6).
Poncy, O., 14 (1-14).
Prévost, M.F., *et al.*, 559, 680a (1-5); 680b (1-14); 1149 (1-3); 1466 (1-14); 2129 (1-6); 2131 (1-4); 2469 (1-3); 3209 (1-6); 3672 (1-14); 3971 (1-12); 4452 (1-3).
Puig, H., 12085 (1-4).
Raynal-Roques, A., 20084 (1-6).
Riéra, B., 1465 (1-6).
Sabatier, D., 1017 (1-14).

---

[8] The prefixes B, BC and T indicate number series under Oldeman's name but made by other collectors

Meliaceae

GUYANA

---

[9] Also cited erroneously in the literature as Wachenheim, G.

Focke, H.C., **1166** (3-3).

Geijskes, D.C., s.n. (9-10).

Gonggrijp, J.W., 400 (4-2); 1030 (3-3); 3775 (5-8).

Granville, J.J. de, 11946 (9-12); 12028 (5-8).

Hall, C.J.J. van, 25 (3-3).

Hammel, B., *et al.*, 21298 (5-12); 21388 (9-1); 21404 (9-9); 21416 (9-15); 21527, 21569 (9-1); 21570, 21616 (5-12); 21662 (5-13).

Henker, G.G., 10441 (3-3).

Heyde, N.M., 184 (3-3); 342 (9-10).

Hostmann, F.W.R., 293 (5-8); 347 (9-10); **662** (9-16); **1204a** (9-12).

Irwin, H.S., *et al.*, 54697, 54707 (9-16); 54723 (9-12); 54725, 54752, 54753, 54830, 54992, 55245, 55372 (9-16); 55558 (9-12); 55570 (9-16); 55630 (9-12); 55759 (5-12a); 55855 (9-12); 55890, 57509, 57577 (9-16).

Jansen-Jacobs, M.J., *et al.*, 6492 (9-16); 6722 (5-12).

Jiménez-Saa, J.H., LBB 14254 (3-3); LBB 14269 (5-6).

Jong, B. de, LBB 15787 (9-12).

Jonker, F.P. & A.M.W. Jonker-Verhoef, 242, 518 (5-8); 581 (9-10).

Kanhai, E.D., LBB 13290 (5-9); LBB 13308 (9-8).

Kappler, A., **2014** (9-14); **2130** (9-7).

Kramer, K.U., 2567, 2762 (5-8); 2788, 2990 (9-10).

Lanjouw, J., *et al.*, 90 (5-8); 409 (3-3); 833 (5-9); 835 (9-12); 836 (5-7); 1159, 1163 (5-8); 1350 (9-10); 1364, 1377, 1420 (9-10); 1464, 1696 (5-8); 1802 (9-10); 1882, 2057 (5-8); 2277, 2500 (5-4); 2563 (5-12a); 2584 (9-8); 2900 (5-7).

LBB (Lands Bosbeheer), see also Boerboom, Borsboom, Elburg, Jiménez-Saa, de Jong, Kanhai, Maas, Reeder, Roberts, Schulz, Teunissen, Tjon Lim Sang, van Troon.

Lems, K., 5092 (4-2); 5075 (3-3).

Lindeman, J.C. *et al.*, 1964 (3-2); 3618 (5-9); 3636 (9-16); 3649 (5-9); 3761 (5-8); 4077 (9-16); 4121 (9-8); 4578 (9-14); 4744, 4902 (5-12a); 4934 (3-3); 5045 (5-9); 5075, 5262 (9-12); 5486 (5-8); 5499 (9-10); 5521 (5-8); 5819 (9-8); 6286 (9-7); 6287 (9-10); 6316, 6331 (9-7); 6331a (9-16); 6395 (9-8); 6738 (9-16); 6758 (9-8); 6787 (5-6); 6977 (9-8).

Lindeman, J.C. & A.R.A. Görts-van Rijn *et al.*, 510 (9-16); 515 (9-8).

Lindeman, J.C. & E.A. Mennega *et al.*, 75 (9-12).

Lindeman, J.C. & A.C. de Roon *et al.*, 830 (9-6); 831 (5-8); 890 (5-7).

Lindeman, J.C. & A.L. Stoffers *et al.*, 97, 130 (5-4); 193 (9-16); 236 (5-4); 534 (9-16); 750 (9-6).

Maas, P.J.M., *et al.*, LBB 10724 (5-9); LBB 10796 (5-12a); LBB 10841 (9-6); LBB 11022 (9-12); LBB 11024 (5-9); LBB 11025 (5-6); LBB 11046 (9-8).

Maguire, B., 24151 (5-12a); 24182, 24195 (9-14); 24333 (9-8).

Mennega, A.M.W., 189 (5-8); 207 (9-10); 457 (5-8); 480 (9-7); 522 (9-12); 523 (9-7).

Oldenburger, F.H.F., *et al.*, 432 (9-10); 942 (5-8); 1111 (5-1); 1181 (5-12a); 1201 (5-4); 1254 (5-1).

Outer, R.W. den, 936 (9-16); 865, 908 (5-8).

Palmer-Jones, R.W., 73 (5-8).

Pulle, A.A., 25 (3-3); 133 (5-8); 135 (9-14); 152 (3-3); 211 (9-16); 324 (5-6); 380 (5-9).

Reeder, D., LBB 12314 (5-12a).

Reijenga, T.W., 561, 698 (5-8).

158

Roberts, L., LBB 16283 (5-5).

Samuels, J.A., 222 (9-8); 530 (9-15); 533 (5-8); s.n. (9-15).

Schulz, J.P., 7192 (4-2); 7203 (9-8); 7452 (5-9); 7502, 7627, 7666, 7719 (9-8); 7726 (9-16); 7758, 7814 (9-15); 7849 (9-12); 8019 (9-8); LBB 8110 (9-10); LBB 8151 (5-13); LBB 8230, LBB 8390 (5-9); LBB 8570 (9-16); LBB 8590 (5-12a); LBB 8983 (5-6); LBB 9001 (3-2); LBB 9002, LBB 9317 (3-3); LBB 10196 (5-9).

Splitgerber, F.L., **311** (9-10).

Stahel, G., 105? (3-3); 167 (4-2); 182 (9-14); 184, 184a (5-8); 242 (5-12a); 362 (5-9); 365 (9-7).

Teunissen, P.A., et al., LBB 15055 (5-8); LBB 15144 (9-10); LBB 15298 (9-16).

Tjon Lim Sang, R.J.M., LBB 16224 (5-7).

Tresling, J.H.A.T., **281** (9-4).

Troon, F. van, LBB 15224 (5-10a); 16308 (5-9).

Versteeg, G.M., 766 (5-12a).

Wessels Boer, J.G., 1188, 1287 (5-12a).

Woodherbarium Suriname, see Stahel.

Wullschlägel, H.R., 61 (5-8); **1333** (9-16).

Zanderi, J., 297 (9-8).

FRENCH GUIANA

Aublet, J.B.C.F. d', **s.n.** (3-2).

Acevedo-Rodriguez, P., et al., 4978 (5-12); 11093 (9-14).

Avril, C., 33 (9-8).

Barrabé, L., 218 (1-1)

Barrier, S., 2599, 3808, 4246 (9-12); 4891 (9-15).

Béna, P., 1048, 1160, 1170, 1173 (3-2).

Benoist, R., 1265 (9-14).

Billiet, F. & B. Jadin, 1011 (5-8).

Boom, B.M., et al., 1753 (9-10); 1833 (3-3); 1845 (9-8); 1892 (9-10); 1947 (5-6); 2012, 2070 (5-9); 2099 (5-6); 2100, 2105 (5-9); 2119 (5-7); 2133 (5-6); 2154 (3-3); 2188, 2196 (5-9); 2215, 2219 (5-6); 2254 (5-9); 2269 (5-7); 2286 (3-2); 2382 (3-3); 10825 (9-4).

Broadway, W.E., 226 (8-1); 393 (5-8).

Champagne, H., 45 (3-2).

Collector unknown, **s.n.** (5-7); **s.n.** (5-8); **s.n.** (5-12a); **s.n.** (5-13); **s.n.** (9-2).

Cremers, G., et al., 4505 (5-12b); 5000 (9-2); 5029 (3-3); 6491 (9-16); 7345 (5-8); 7726 (5-12); 8405 (5-8); 11113 (5-12a); 11868, 11959 (5-11).

Delnatte, C., 1479 (5-11).

Ek, R.C., et al., 1286 (3-2); 1312 (5-11).

Engel, J. & M. Tarcy, 4 (5-10a).

Feuillet, C., et al., 46 (5-6); 1584, 2328 (5-8); 9983, 10069 (5-12).

Fleury, M., 300 (9-14); 522, 841 (5-8); 1365, 1631 (5-12).

Gentry, A., 50289 (3-2).

Granville, J.J. de, et al., 650 (5-12a); 985 (5-12); T-1093 (3-2); 1220 (2-1); 1769 (5-12); 1867 (9-16); 1987, 2113 (5-12); 2492 (5-9); B-3633 (9-16); B-3756 (5-12a); 4393, 4447 (5-12); B-4690 (5-10); 4719 (5-12); B-4791 (5-12a); B-4946 (9-8); 5650 (9-15); 6589 (9-16); 6753 (9-8); 7519 (5-12a); 7566 (5-13); 7628 (5-7); 7652 (5-4); 7749 (5-11); 7876 (5-12a); 7971 (9-16); 8065 (5-8); 8072 (5-4); 8121 (5-12); 8821 (5-11); 9172 (5-12); 9181 (9-15); 9230, 9607 (5-12); 9843 (5-6); 10334 (5-12); 10707 (5-4); 11170 (5-11); 11893 (9-16); 11946 (9-12); 11962, 12008 (5-12); 12122 (9-2); 12351, 12904 (5-12); 15977 (5-13).

Sastre, C., *et al*., 1709 (9-8); 3899 (5-12a); 3952, 5642 (5-12); 5905 (5-6); 6209 (3-3); 6298 (9-16); 8139 (9-16); 8152 (5-12a).

Schnell, R., 11871 (5-12).

Service Forestier Guyane Française (= BAFOG), 177M (3-3); 184, 184a (5-8); 353M (9-14); 1069 (5-8); 1215, 4465 (3-3); 4832 (5-6); 6097 (3-2), 7074, 7096, 7331, 7365 (3-3); 7755 (5-12a).

Smith, N.P., *et al*., 13 (5-14).

Vieillescazes, A., 438, 441, 479 (9-16); 519 (5-9).

Vigneron, M., 19 (7-1).

Villiers, J.-F., 1852 (5-6); 1941 (5-14); 1945a (9-15); 2007 (3-2); 2031 (9-15); 2054 (5-14); 2105, 2156, 2161, 2170 (9-15); 2190 (5-12); 2741 (9-12).

Wachenheim, H., 288, 297, 347 (9-14); 350 (9-16); s.n. (3-3).

Wallnöfer, B., 13497 (5-9).

INDEX TO SYNONYMS AND NAMES IN NOTES

rotundifolia = 6-2b
surinamensis var. reticulata = 6-7
trinitensis = 6-2

Tigarea
 aspera = 2-2, 6-5
 dentata = 3-3

Vitaceae

Cissus
 boliviana Lombardi = 1-10
 duarteana Cambess., see 1-5, note
 elongata Miq. = 1-5
 erosa Rich., see 1-4, note
 erosa Rich. subsp. *linearifolia* (Baker) Lombardi= 1-5, and note
 erosa Rich. var. *linearis* (Baker) Hoehne = 1-5
 erosa Rich. var. *salutaris* (Kunth) Planch. = 1-5
 flavens Desc. = 1-2
 guyanensis Desc.= 1-3, and note
 kawensis Desc. = 1-3
 lucida Poir. = 1-5
 obscura DC. = 1-14
 ovata Rich. = 1-14
 nobilis Kuhlm., see 1-11, note
 parkeri (Baker) Planch. = 1-8
 peruviana Lombardi, see 1-12, note
 puncticulosa Rich. = 1-14
 rhombifolia Vahl = 1-1
 salutaris Kunth = 1-5, and note
 selloana (Baker) Planch. = 1-10
 sicyoides L. = 1-14
 subrhomboidea (Baker) Planch., see 1-4, note
 trigona Willd. ex Schult. & Schult. f., see 1-1, note
 verticillata (L.) Nicolson & C.E. Jarvis
  subsp. colombiana Lombardi, see 1-14, note
  subsp. micrantha (Poir.) Lombardi, see 1-14, note
  subsp. oblongolanceolata (Krug & Urb.) Lombardi, see 1-14, note
 vitiginea L., see 1, type

Phoradendron
 verticillatum (L.) Druce = 1-14

Viscum
 verticillatum L. = 1-14

Vitis

alata (Jacq.) Kuntze = 1-1
duarteana (Cambess.) Baker = 1-4
erosa (Rich.) Baker = 1-5
miqueliana Baker = 1-5
miqueliana Baker var. *linearifolia* Baker = 1-5
parkeri Baker = 1-8, and note
rhombifolia (Vahl) Baker = 1-1
salutaris (Kunth) Baker = 1-5
selloana Baker = 1-10
sicyoides (L.) Morales = 1-14
spinosa (Cambess.) Baker = 1-8
ulmifolia Baker = 1-12
vitiginea (L.) W.L. Theob. var. *sicyoides* (L.) Kuntze = 1-14

Meliaceae

Cabralea
    canjerana subsp. polytricha (A. Juss.) T.D. Penn.,
        see 2, type and 2-1, note
    polytricha A. Juss., see 2, type

Carapa
    procera DC., see 3-3, note

Cedrela
    mahagoni L., see 8, type

Granatum
    surinamensis (Miq.) Kuntze = 3-3

Guarea
    affinis A. Juss. = 5-12a
    aubletii A. Juss. = 5-8
    concinna Sandwith = 5-12a
    cristata T.D. Penn., see 5-11, note
    davisii Sandwith = 5-12a
    macrophylla Vahl subsp. pachycarpa (C. DC.) T.D. Penn., see 5-6, note
    megantha A. Juss. = 5-7
    pendulispica C. DC. = 5-10b
    pubiflora A. Juss. = 5-12b
    richardiana A. Juss. = 5-12a
    schomburgkii C. DC. = 5-5
    trichilioides L., see 5, type
    trichilioides L. var. *pachycarpa* C. DC. = 5-10a

Moschoxylum
    cipo A. Juss. = 9-2

pleeanum A. Juss. = 9-11
propinquum Miq. = 9-12
surinamense Miq. = 9-16

Odontandra
    quadrijuga (Humb., Bonpl. & Kunth) Triana & Planch = 9-12

Portesia
    echinocarpa de Vriese = 9-10

Samyda
    guidonia L. = 5-8

Swietenia
    mahagoni (L.), see 8, type
    senegalensis Desr. = 6-1

Trichilia
    acariaeantha Harms = 9-8
    alternans C. DC. = 9-16
    brachystachya Klotsch ex C. DC. = 9-10
    canjerana Vell. = 2-1
    compacta A.C. Sm. = 9-12
    cuneifolia Pulle = 9-14
    davisii Sandwith = 9-10
    echinocarpa (de Vriese) Walp. = 9-10
    elegans A. Juss. subsp. richardiana (A. Juss.) T.D. Penn., see 9-3, note
    fuscescens Radlk. = 9-7
    guianensis Klotsch ex C. DC. = 9-13
    guianensis Klotsch ex C. DC. var. *parvifolia* C. DC = 9-13
    lepidota Mart. subsp. lepidota, see 9-6, note
    lepidota Mart. subsp. schumanniana (Harms) T.D. Penn., see 9-6, note
    leucastera Sandwith = 9-6
    moritzii C. DC. = 9-15
    propinqua (Miq.) C. DC. = 9-12
    pubescens Rich. = 5-12
    quadrijuga Kunth subsp. cinerascens (C. DC.) T.D. Penn., see 9-12, note
    roraimana C. DC. = 9-8
    schomburgkii C. DC. subsp. javariense T.D. Penn., see 9-14, note
    stelligera Radlk. = 9-4
    subsessilifolia C. DC. = 9-14

INDEX TO VERNACULAR AND TRADE NAMES

Alphabetic list of families of series A occurring in the Guianas

Defined as in Cronquist, 1981, and numbered in his sequence, with alternative names. Those published, with chronological fascicle number and year.

Abolbodaceae			Cactaceae	031	18. 1997	
(see Xyridaceae	182)	15. 1994	Caesalpiniaceae	088	p.p.	7.
Acanthaceae	156	23. 2006			1989	
(incl. Thunbergiaceae)			Callitrichaceae	150		
(excl. Mendonciaceae	159)		Campanulaceae	162		
Achatocarpaceae	028	22. 2003	(incl. Lobeliaceae)			
Agavaceae	202		Cannaceae	195	1. 1985	
Aizoaceae	030	22. 2003	Canellaceae	004		
(excl. Molluginaceae	036)	22. 2003	Capparaceae	067		
Alismataceae	168	27. 2009	Caprifoliaceae	164		
Amaranthaceae	033	22. 2003	Caricaceae	063		
Amaryllidaceae			Caryocaraceae	042		
(see Liliaceae	199)		Caryophyllaceae	037	22. 2003	
Anacardiaceae	129	19. 1997	Casuarinaceae	026	11. 1992	
Anisophylleaceae	082		Cecropiaceae	022	11. 1992	
Annonaceae	002		Celastraceae	109		
Apiaceae	137		Ceratophyllaceae	014		
Apocynaceae	140		Chenopodiaceae	032	22. 2003	
Aquifoliaceae	111		Chloranthaceae	008	24. 2007	
Araceae	178		Chrysobalanaceae	085	2. 1986	
Araliaceae	136		Clethraceae	072		
Arecaceae	175		Clusiaceae	047		
Aristolochiaceae	010	20. 1998	(incl. Hypericaceae)			
Asclepiadaceae	141		Cochlospermaceae			
Asteraceae	166		(see Bixaceae	059)		
Avicenniaceae			Combretaceae	100	27. 2009	
(see Verbenaceae	148)	4. 1988	Commelinaceae	180		
Balanophoraceae	107	14. 1993	Compositae			
Basellaceae	035	22. 2003	(= Asteraceae	166)		
Bataceae	070		Connaraceae	081		
Begoniaceae	065		Convolvulaceae	143		
Berberidaceae	016		(excl. Cuscutaceae	144)		
Bignoniaceae	158		Costaceae	194	1. 1985	
Bixaceae	059		Crassulaceae	083		
(incl. Cochlospermaceae)			Cruciferae			
Bombacaceae	051		(= Brassicaceae	068)		
Bonnetiaceae			Cucurbitaceae	064		
(see Theaceae	043)		Cunoniaceae	081a		
Boraginaceae	147		Cuscutaceae	144		
Brassicaceae	068		Cycadaceae	208	9. 1991	
Bromeliaceae	189	p.p. 3. 1987	Cyclanthaceae	176		
Burmanniaceae	206	6. 1989	Cyperaceae	186		
Burseraceae	128		Cyrillaceae	071	27. 2009	
Butomaceae			Dichapetalaceae	113	27. 2009	
(see Limnocharitaceae	167)	27. 2009	Dilleniaceae	040	31. 2016	
Buxaceae	115a		Dioscoreaceae	205		
Byttneriaceae			Dipterocarpaceae	041a	17. 1995	
(see Sterculiaceae	050)		Droseraceae	055	22. 2003	
Cabombaceae	013		Ebenaceae	075		

168

Elaeocarpaceae	048	
Elatinaceae	046	
Eremolepidaceae	105a	25. 2007
Ericaceae	073	
Eriocaulaceae	184	
Erythroxylaceae	118	
Euphorbiaceae	115	
Euphroniaceae	123a	21. 1998
Fabaceae 089		
Flacourtiaceae	056	
(excl. Lacistemaceae	057)	
(excl. Peridiscaceae	058)	
Gentianaceae	139	30. 2014
Gesneriaceae	155	26. 2008
Gnetaceae	209	9. 1991
Gramineae		
(= Poaceae	187)	8. 1990
Gunneraceae	093	
Guttiferae		
(= Clusiaceae	047)	
Haemodoraceae	198	15. 1994
Haloragaceae	092	
Heliconiaceae	191	1. 1985
Henriquesiaceae		
(see Rubiaceae	163)	
Hernandiaceae	007	24. 2007
Hippocrateaceae	110	16. 1994
Humiriaceae	119	
Hydrocharitaceae	169	
Hydrophyllaceae	146	
Icacinaceae	112	16. 1994
Hypericaceae		
(see Clusiaceae	047)	
Iridaceae	200	
Ixonanthaceae	120	
Juglandaceae	024	
Juncaginaceae	170	
Krameriaceae	126	21. 1998
Labiatae		
(= Lamiaceae	149)	
Lacistemaceae	057	
Lamiaceae	149	
Lauraceae	006	
Lecythidaceae	053	12. 1993
Leguminosae		
(= Mimosaceae	087)	
+ Caesalpiniaceae	088)	p.p. 7. 1989
+ Fabaceae	089)	
Lemnaceae	179	
Lentibulariaceae	160	
Lepidobotryaceae	134a	
Liliaceae 199		
(incl. Amaryllidaceae)		
(excl. Agavaceae	202)	
(excl. Smilacaceae	204)	
Limnocharitaceae	167	27. 2009
(incl. Butomaceae)		

Linaceae	121	
Lissocarpaceae	077	
Loasaceae	066	
Lobeliaceae		
(see Campanulaceae	162)	
Loganiaceae	138	
Loranthaceae	105b	25. 2007
Lythraceae	094	
Malpighiaceae	122	
Malvaceae	052	
Marantaceae	196	
Marcgraviaceae	044	
Martyniaceae		
Mayacaceae	183	
Melastomataceae	099	13. 1993
Meliaceae	131	31. 2016
Mendonciaceae	159	23. 2006
Menispermaceae	017	
Menyanthaceae	145	
Mimosaceae	087	28. 2011
Molluginaceae	036	22. 2003
Monimiaceae	005	
Moraceae	021	11. 1992
Moringaceae	069	
Musaceae	192	1. 1985
(excl. Strelitziaceae	190)	
(excl. Heliconiaceae	191)	
Myoporaceae	154	
Myricaceae	025	
Myristicaceae	003	
Myrsinaceae	080	
Myrtaceae	096	
Najadaceae	173	
Nelumbonaceae	011	
Nyctaginaceae	029	22. 2003
Nymphaeaceae	012	
(excl. Nelumbonaceae	010)	
(excl. Cabombaceae	013)	
Ochnaceae	041	
Olacaceae	102	14. 1993
Oleaceae	152	
Onagraceae	098	10. 1991
Opiliaceae	103	14. 1993
Orchidaceae	207	
Oxalidaceae	134	
Palmae		
(= Arecaceae	175)	
Pandanaceae	177	
Papaveraceae	019	
Papilionaceae		
(= Fabaceae	089)	
Passifloraceae	062	
Pedaliaceae	157	
(incl. Martyniaceae)		
Peridiscaceae	058	
Phytolaccaceae	027	22. 2003
Pinaceae	210	9. 1991

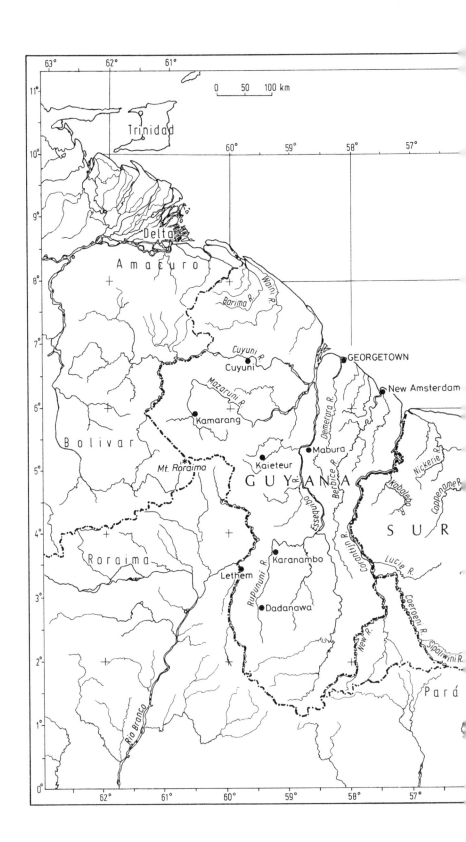

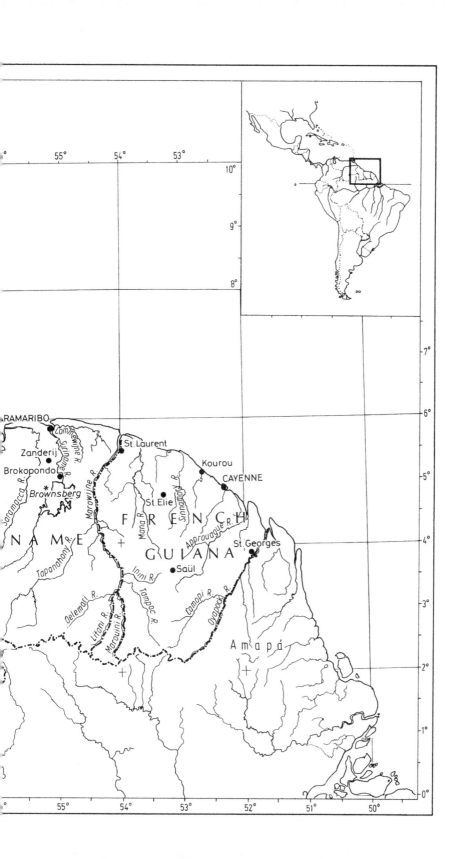

Printed by Printforce, United Kingdom